Praise for *The Brain Under Siege*

"If you're dealing with the diagnosis—either for yourself or a loved one—of a neurologic condition, you're likely feeling lost and overwhelmed. Look no further than this book, which not only breaks down what's happening in the brain like a crime scene, but inspires hope for future treatments. It's hard to get more complex than the human brain. In *The Brain Under Siege*, Howard L. Weiner has created a remarkable guide to neurologic diseases that we all can understand."

—Ann Romney

"Howard Weiner has an amazing ability to synthesize complex medical and scientific concepts into something accessible and understandable. He has been on the front lines of brain disease research for over 40 years. In *The Brain Under Siege* he tells the story of five of the most devastating neurological diseases. He likens each disease to a crime scene and explains how the crime will ultimately be solved. Not only is this book highly informative, it's also genuinely fun to read."

—Stephen L. Hauser, MD, Robert A. Fishman Distinguished Professor of Neurology at the University of California, San Francisco, and director of the UCSF Weill Institute for Neurosciences

"Howard Weiner's latest book, *The Brain Under Siege*, breaks down the complexities of brain health and disease in an accessible and enjoyable way. He opens up the world of drug development and how scientific discovery moves from the laboratory to clinical trials and ultimately to FDA approval."

—Al Sandrock, MD, PhD, head of research and development at Biogen

"A fascinating read, *The Brain Under Siege* is educational, memorable, and perhaps most important, optimistic. Chances are you know someone affected by one of the common illnesses of the brain. *The Brain Under Siege* explains the science behind these diseases and the possible pathways to a cure. Dr. Howard L. Weiner has done an inspiring job of putting it all together. This is a valuable resource for both the medical community and the lay public."

—Merit Cudkowicz, MD, Julieanne Dorn Professor of Neurology at Harvard Medical School, chief of neurology and director at Healey ALS Center, Massachusetts General Hospital

"The stakes are high when the brain is in trouble. Howard L. Weiner chronicles the heroic efforts that have led to astounding progress in our ability to understand and treat the most essential and most fragile part of our humanity. For each brain disease explored, the analogy of a crime scene draws the reader into the detective work that goes on as researchers painstakingly track down suspected causes. The *Brain Under Siege* reveals the drama behind the science as doctors grapple with the most complicated structure in the universe."

—Daniel Z. Lieberman, MD, coauthor of *The Molecule of More*

"When it comes to complex medical issues, patients and their families are usually overwhelmed. The situation is complex even for scientists in the field. In Howard Weiner's ambitious new book, *The Brain Under Siege,* this talented neurologist tackles brain disease and unravels its complexity—no easy feat for such a heavy subject. Likening the diseased brain to a crime scene, he takes the reader on the challenging journeys to find the elusive cures for these brain diseases. The book is a must read for anyone who wants to learn more about the scientific process and the urgent quest to treat—and then prevent—chronic brain diseases."

—Dennis J. Selkoe, MD, Vincent and Stella Coates Professor of Neurology at Harvard Medical School and codirector of the Ann Romney Center for Neurologic Diseases at Brigham and Women's Hospital

ALSO BY HOWARD L. WEINER, MD

Nonfiction

Neurology for the House Officer

Curing MS

Fiction

The Children's Ward

Film

What is Life? The Movie

Abe and Phil's Last Poker Game

The Brain Under Siege

Solving the Mystery of Brain Disease, and How Scientists Are Following the Clues to a Cure

HOWARD L. WEINER, MD

BenBella Books, Inc.
Dallas, TX

BENBELLA

BenBella Books, Inc.
10440 N. Central Expressway
Suite 800
Dallas, TX 75231
benbellabooks.com
Send feedback to feedback@benbellabooks.com

Printed in the United States of America
10 9 8 7 6 5 4 3 2 1

Library of Congress Control Number: 2021012421
ISBN 9781953295545 (hardback)
ISBN 9781953295880 (ebook)

Editing by Sheila Curry Oates
Copyediting by Judy Myers
Proofreading by Greg Teague and Cape Cod Compositors, Inc.
Indexing by WordCo Indexing Services, Inc.
Text design and composition by PerfecType, Nashville, TN
Cover design by Devin Watson
Printed by Lake Book Manufacturing

To Dennis Selkoe—for sharing the journey

and

To Ann Romney—for her courage and inspiration

CONTENTS

CHAPTER ONE

The Brain Under Siege

In 1997, I was in the middle of a lab meeting when I received an urgent call from Dr. Gordon Williams, head of endocrinology at the Brigham and Women's Hospital. Gordon and I had worked together on a number of scientific projects over the years. Gordon's call was about a woman, a friend of his, who had just been diagnosed with multiple sclerosis. She was in distress, and he asked whether I could see her immediately. I agreed and soon found myself sitting across from a handsome couple, who were holding hands as the woman told me her story. She was forty-eight years old, and her symptoms began with numbness, fatigue, and lack of coordination. She felt that she was losing her ability to live her life, losing the ability to be the person, the mother, the wife that she had always been. When she was diagnosed, she was told there was little that could be done and to come back when her symptoms got worse. So she came to see me for a second opinion.

Whenever I see a patient who comes in with a diagnosis of MS, my first task is to make sure that the diagnosis is indeed correct. More than once, I have immediately "cured" MS when I discovered there had been an incorrect diagnosis. In this case, however, the diagnosis was unequivocal:

the exam, the history, and an MRI showed abnormalities in the brain and spinal cord that all fit with MS. I was surprised that no treatment had been suggested and told the patient that there were in fact things we could do, and not to give up hope. I knew that more than anything, she needed hope.

I explained to her and her husband that we needed to treat her immediately, took her by the hand, and led her to the Infusion Center, where we began to give her large doses of corticosteroids to dampen the inflammation in her brain. She was seated next to people in wheelchairs, some getting chemotherapy, but much of the fear and uncertainty that she had shown when I began my exam had dissipated, because something was being done. I never reveal the identity of a patient, except in this case the patient has already told her story publicly. The patient was Ann Romney, and she has become a crucial advocate in the search for treatments and cures for debilitating neurological diseases. She and her husband, Mitt, have led the charge in raising philanthropic funds for the Ann Romney Center for Neurologic Diseases at Brigham and Women's Hospital, where Dennis Selkoe and I have devoted our lives to cross-pollinating research across MS and Alzheimer's. With the formation of the Ann Romney Center, we raised our efforts to a higher level and added amyotrophic lateral sclerosis (ALS), Parkinson's disease, and brain tumors to our research goals.

In this book, I hope to tell you the story of that research and the amazing strides we are making in searching for cures for these neurological diseases—but also how far we have to go. Above all, I will tell you what Dennis and I told Ann Romney: "There is hope, now more than ever."

ABOUT THE BRAIN

How should we think about the brain? It is not an easy question. It's much easier to think about the heart or the kidneys. The heart is a pump that circulates blood throughout the body. The kidneys are filters that cleanse the blood. The brain is more complex and is what separates us from all other life on earth. To ask, How does one think about the brain? creates the conundrum of the brain thinking about itself. Sidestepping the philosophical question of the brain contemplating itself, the healthy brain has a vast number of functions.

First, the brain serves as a center that controls movement. If I want to move my hand, a nerve center in one particular area of the brain, called Broca's motor area, fires. The impulse travels in nerve bundles that cross from one side of the brain to the other, enter the spinal cord, then go out of the spinal cord to the nerves of the hand and to the muscles that manipulate my hand. Because the fibers cross from one side of the brain to the other, the right side of the brain controls the left side of the body. Accordingly, a stroke on the right side of the brain causes paralysis on the left side of the body. The brain not only controls movement but also helps manage breathing, blood pressure, heartbeat, and digestion.

Second, in addition to sending out impulses, the brain receives impulses. All sensory input from the environment congregates in the brain: vision, hearing, smell, touch, and taste. Each has its own special pathway, and the sensory fibers also cross before they get to the thinking part of the brain. If I touch something hot, fibers transmit the pain to the brain, which then triggers motor fibers, which in turn move my hand away from the hot object. Sometimes no connection within the brain is needed to generate movement. If I tap my knee with my reflex hammer, a connection is made in my spinal cord and my leg jerks, an involuntary movement. I am not in control.

Third, the brain not only functions to control movement and receive sensory input but it is also the seat of consciousness. We know consciousness exists, we feel it, but we still don't understand it.

Finally, the brain is a computer, albeit an imperfect one. The brain's wondrous complexity, however, has a flip side: it is one of the most difficult organs to study when it is affected by disease.

This book will focus on five brain diseases that have long frustrated researchers and curtailed lives: multiple sclerosis (MS), Alzheimer's disease, Parkinson's disease, amyotrophic lateral sclerosis (ALS) also known as Lou Gehrig's disease, and glioblastoma brain tumors. In this book, I will tell the stories of that research, and the amazing strides the scientific community is making in developing therapies for some of these chronic neurological diseases—but also how far we have to go with others.

The brain is the last great frontier of medical science. For decades, we have made extraordinary strides in extending and improving human life. In treating cardiovascular disorders, for example, we are a world away from where we were in the 1950s, when statins were not yet invented and heart surgery was a blunt, brutal affair. Now, medical science has made cardiac surgery incredibly precise and almost routine. The mortality rate for open-heart surgery was 50 percent half a century ago. It's now less than 2 percent. It's amazing to think how far we have come. But for so long, the mysteries of brain disease have eluded us.

Today, researchers around the world are finally making some of the most important strides we've seen in our understanding of the brain—and not a moment too soon, as the World Health Organization (WHO) reports that up to a billion people around the world suffer from some sort of neurological condition. Millions suffer from some of the most debilitating forms of those diseases: MS, ALS, Alzheimer's, and Parkinson's. These are the diseases that, as many of my patients have told me, rob people of their very essence.

The brain is made up of some two hundred billion cells, a staggering number, and there are four major types of cells in the brain. Perhaps the most familiar to nonscientists are the nerve fibers, or neurons. Neurons have long, specialized extensions that transmit electrical impulses and carry signals from one part of the brain to another, from the brain to all other parts of our body, and from other parts of the body back to the brain. These neurons can be thought of as electrical wires, and they are covered by an insulation called the myelin sheath—this is the structure that erodes in multiple sclerosis.

The second type of cell in the brain is called the oligodendrocyte. It wraps around the neuron in multiple concentric circles and creates the myelin sheath, which is the insulation surrounding the neuron that

THE NORMAL BRAIN

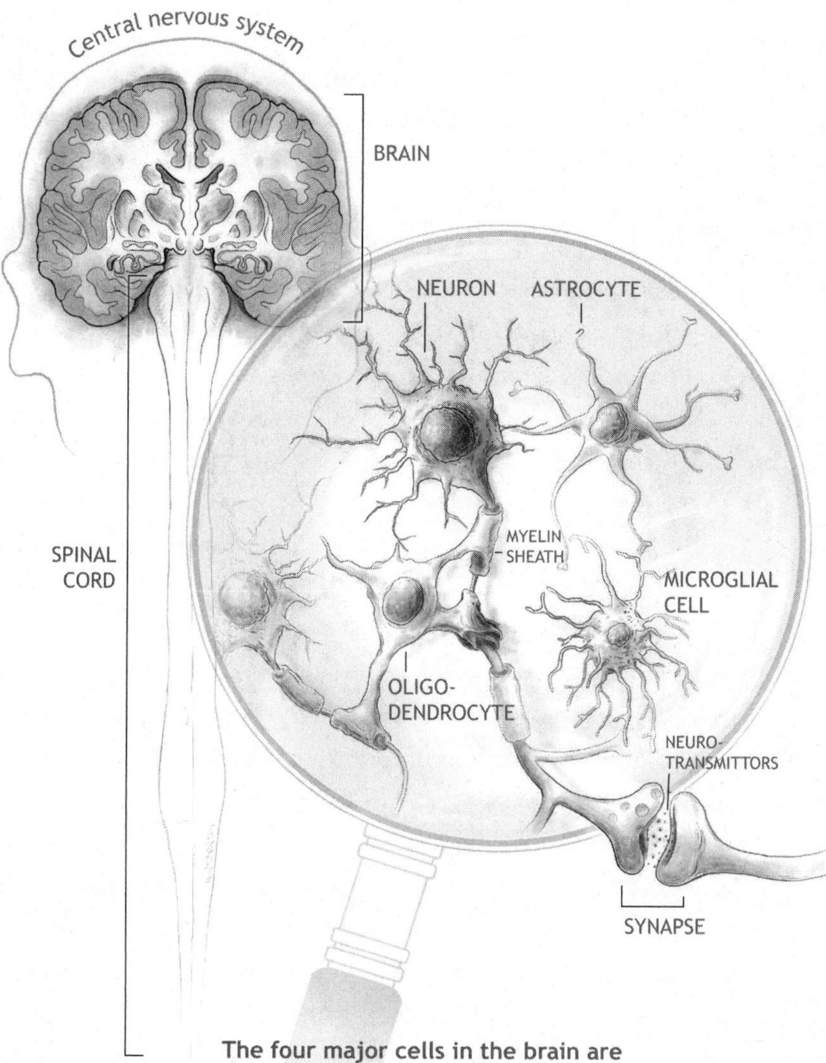

The four major cells in the brain are neurons, microglial cells, astrocytes, and oligodendrocytes that wrap around neurons to form the myelin sheath.

accelerates nerve conduction. Electricity that moves across the neuron does not travel in a straight line on the nerve itself; it jumps from one place on the nerve to another where there is a break in the insulation of the myelin sheath that was created by the oligodendrocyte. These breaks are called the nodes of Ranvier. Here's a way to think about how these breaks in the myelin sheath work: Imagine traveling from New York to California. One way is to drive cross-country on the interstate highways. If you only stop to sleep along the way, you can make it in four days. Of course, it's quicker to fly. If you take a plane that touches down briefly on its journey, first in Chicago, then in Denver, and ultimately arriving in California, it will only take hours (assuming, of course, there is no snow in Chicago or Denver). That's how electricity travels along the nerve fibers in our brain. The insulation created by the myelin sheath allows the nerve impulses to jump from one break in the insulation to another.

MS is called a demyelinating disease because the insulation on the nerve is attacked by the immune system. Because of damage to the myelin sheath, the neurons are slowed in their ability to conduct nerve impulses, therefore becoming unable to travel from one node to another.

The nervous system is divided into the central nervous system, which consists of the brain and spinal cord, and the peripheral nervous system, which is made up of the nerves that run back and forth from the spinal cord to the muscles and other organs. Myelin surrounds most nerve fibers in both the central and peripheral nervous systems. The equivalent of the oligodendrocyte in the peripheral nervous system is called the Schwann cell. Interestingly, the peripheral nervous system is much better at being able to remyelinate or rebuild damaged myelin, which poses a major challenge in treating MS: how to rebuild damaged myelin in the central nervous system. We do not yet have that answer, but in chapter two, I will explore some of the exciting approaches and experimental therapies that are currently being investigated.

Back to the basics of brain cells. In addition to the neuron and the oligodendrocyte, the third major cell type in the brain is called a microglial cell. Microglial cells comprise the brain's own internal immune system. They are the brain's sentinels that respond to infection, trauma, or any insult that occurs in the brain. The microglial cell is also a gardener in the brain and prunes the nerve fibers to keep them in good shape. They can remove toxic substances that accumulate throughout the brain. Microglial cells are normally in a resting state until triggered by either external or internal changes. They then become activated and change their shape as they attempt to remedy whatever has provoked them into action, whether it is an infection or damage caused by trauma to the brain. As we get older, microglial cells do not function as well, contributing to the accumulation of toxic proteins from diseased nerve cells in the brain. Indeed, microglial cells themselves can become toxic and cause damage, and they do so in diseases such as ALS, Alzheimer's disease, and Parkinson's disease. It may be that their loss of function when we get older contributes to the accumulation of the toxic protein beta-amyloid seen in the brains of people with Alzheimer's disease.

The fourth major cell in the brain is called an astrocyte, which supports brain structure and helps form the blood-brain barrier. The blood-brain barrier is a network of blood vessels that creates a boundary that prevents certain substances from entering the brain, including some drugs. Francisco Quintana in our center has shown that astrocytes communicate with microglial cells and neurons. He also found that there are subtypes of astrocytes that can protect or harm brain function depending on the genes they express. Astrocytes can also grow out of control; when they do, they become malignant and form brain tumors.

Although there are others, the work these four major cells of the brain do is astounding. The one hundred billion neurons in the brain are each believed to be connected—on average—to some ten thousand other cells, which create almost one thousand trillion connections in the brain: more than the stars in the Milky Way. Like the incredibly complex supercomputer that it is likened to, rapid communication in the brain is essential to its function. Neurons communicate with other neurons by releasing chemicals, or neurotransmitters, at the nerve terminal. Once released, a neurotransmitter moves to another neuron and delivers its message. The communication and connection are similar to a runner handing a baton to another runner in a race. Neurotransmitters can be excitatory and stimulate the nerve, while other neurotransmitters are inhibitory and inhibit or restrict the function of nerves.

Acetylcholine is an excitatory neurotransmitter that is needed to form and retrieve memories. In Alzheimer's disease, patients suffer from a shortage of acetylcholine in certain parts of the brain that are important for memory. One of the drugs for Alzheimer's disease that has been approved by the Federal Drug Administration (FDA) is Aricept, which prevents the breakdown of acetylcholine in the brain and effectively raises levels of the neurotransmitter. Unfortunately, Aricept only reduces symptoms and does not change the underlying biologic mechanisms that make Alzheimer's a progressive and debilitating disease. We will explore why it has been so difficult to find a treatment for Alzheimer's disease.

Two other neurotransmitters in the brain are GABA and dopamine. GABA is an inhibitory neurotransmitter that serves as a brake on the nervous system so it doesn't get out of control. Drugs that increase GABA levels in the brain are used to treat epilepsy. Dopamine is both an inhibitory and excitatory neurotransmitter that is involved in the control of movement and mood. The loss of dopamine contributes to the slowness of movement and muscle rigidity of Parkinson's disease. Drugs that replace the missing dopamine such as L-DOPA help Parkinson's symptoms but don't impact disease progression. Dopamine is complex in its function: Too much dopamine can cause psychosis, and stimulation of dopamine release is associated with drug addiction.

The brain functions both to control movement and receive sensory input. It is the seat of consciousness, of our awareness of ourselves, and the world around us. The human brain is the most complex biologic system that has arisen on this planet. It is this complexity that makes it arguably the most difficult organ to study—both when it is functioning properly and when it is affected by disease.

Because diseases of the brain are so common, you may be reading this book because of personal experience with one of the major neurological diseases. You may have a relative with MS or a parent or grandparent struggling with Alzheimer's disease. You could be wondering why there have been so many failures in the quest to find treatments for these diseases. You may not be aware of how the diseases interrelate, and how the mechanisms and treatments for one brain disease may impact other neurological diseases of the brain. For decades, researchers and physicians studying the diseases discussed in this book have been relatively siloed—MS research has progressed down one path, and ALS and Alzheimer's disease down their own. But the areas of inquiry that may have seemed peripheral or irrelevant to one particular disease could have huge implications for another. If you are reading this book because you have a relative with one of these diseases, learning about the others as well will provide you with a broader picture of diseases of the brain.

Brain diseases may be classified based on what has gone wrong; what has caused the malfunction. I liken these neurological malfunctions to a crime scene and the search for a range of suspects that could have caused the damage. One of the culprits that is easiest to understand is infection, in which a virus, bacteria, or other microorganism infects the brain. There are many examples: viral encephalitis, HIV/AIDS, and polio. Indeed, in searching for a treatment for a patient with a brain disease, one immediate question is whether there is an infection in the brain. Infections can be treated, and some can be prevented by vaccination.

Another major culprit is indirectly linked to infection and is caused by the immune system. The immune system is the body's defense mechanism. It fights off infection and interacts with infectious agents in our environment. Unfortunately, the immune system can also go astray. For example, in MS, which is now known to be an autoimmune disease, the immune system attacks certain elements of a person's brain and causes damage to the myelin sheath. Interestingly, the immune system can also be harnessed to treat brain diseases, including all the other brain diseases we will discuss—Alzheimer's disease, ALS, Parkinson's disease, and brain tumors.

Genetics may also be a culprit and plays an important role in many brain diseases. If you carry the gene for Huntington's disease, for example, you have an almost 100 percent chance of developing a debilitating dementia. The only way one can get Huntington's is by inheriting this one faulty gene; it cannot be acquired in any other way. There are also genetic forms of Alzheimer's and ALS, wherein several different genes have been identified as culprits.

A very important factor in several progressive, degenerative brain diseases is the accumulation in the brain, as we age, of altered proteins that become toxic and cause brain damage. Two compelling examples are seen in Alzheimer's disease and Parkinson's disease. In Alzheimer's disease, the beta-amyloid protein found outside neurons and the tau protein inside neurons are altered, damaging the neurons and very slowly destroying memory. In Parkinson's disease, a protein called alpha-synuclein accumulates and leads to problems with movement and other symptoms of the disease.

The environment can play a crucial role in brain disease. Interestingly, smoking makes MS worse but decreases the risk for Parkinson's disease. The biggest environmental factor that is only now coming into focus relates to the gut and the trillions of bacteria that reside there, known as the microbiome. As part of the gut-brain axis, these trillions of bacteria have the surprising ability to influence the brain and play a role in each of the five diseases of the brain discussed here.

There are many perpetrators that can lay siege to the brain: an infection, immune cells gone astray that attack the brain, the accumulation of toxic proteins, aberrant genes, and the bacteria in our gut. Understanding the cause of a neurological disorder requires complex detective work that calls for putting together many different clues to understand the crime scene and to devise a treatment that can truly impact the disease.

The brain requires an astounding 25 percent of the body's total blood flow, is dependent on sensory input, and needs sleep. It has tremendous capacity for self-defense, but when faced with some diseases, it may not be able to react properly. The human brain most probably reaches its peak between age twenty and thirty, after which we begin to lose synapses and whole neurons and the brain itself starts to atrophy or shrink. Beginning around age thirty, when we start losing synapses and neurons, our short-term and long-term memories can begin to slip subtly, a process that continues as we age. Can exercise, doing crossword puzzles, or eating the right foods halt or reverse this condition? Probably, but the definitive answer is: we simply don't know. Preventative medicine is a crucial area of research both for the healthy brain and for those who have diseases of the brain. We know that the ultimate cure will likely be prevention.

If one considers embarking on a mission to stop the perpetrators, and to treat a neurological disease, the pathway is easily understood conceptually, although it may be difficult operationally.

1. First, we must understand the workings of a healthy brain.
2. Second, we must identify the exact processes that have gone astray.
3. Third, we must find ways to interrupt or reverse those processes.

Although conceptually straightforward, the reality is that the diagnosis and treatment of neurological conditions is highly complex. Although beyond the scope of this book, part of the healthy brain relates to our very consciousness—and how we define the soul. This remains one of the biggest mysteries in biology. How does the brain create our individual consciousness, our desire for life, our ability to love, our feelings of satisfaction, disappointment, depression, or elation? Certainly, these are questions that physicians have debated for decades and philosophers have explored for centuries.

Because our consciousness and these crucial emotions are what make us human, patients suffering from neurological diseases that affect cognition inevitably are robbed of their very being. A person with a brain tumor may have a change of behavior and personality that resolves when the tumor is removed and returns when the tumor reappears. The soul is even more difficult to define than consciousness. Is the brain the seat of the soul, or does the soul exist independent of the brain? What happens to the soul of a person with Alzheimer's who can no longer recognize his or her family or process sensory information from the environment? These are deeply emotional questions that patients and families struggle with every single day.

When I meet with my colleagues, one of my favorite questions to pose is what I call the "scientific genie" question. It goes like this: If there is one question about MS or Alzheimer's or ALS about which you don't know the answer and there was a genie who could give you the answer, what would you ask? The most common response is: What is the origin of the disease—what is the primary perpetrator? Although our ability to treat neurological diseases is rapidly advancing, this question is central to our understanding of where we go from here.

MAKING REAL PROGRESS

I have spent over forty years caring for people with MS, investigating the disease and struggling to develop new treatments. For all of us who have become older, we find ourselves asking, "Where has the time gone?" For MS, time has not gone to waste. We have made strides with MS that we have yet to make with other neurological diseases. Indeed, the understanding and treatment of MS has been a major success story. We understand the disease better than ever. We know that MS begins as a relapsing disease and then transitions to a more disabling progressive form. We now have over fifteen FDA-approved drugs for relapsing forms of the disease and some drugs now approved for the treatment of progressive forms, although treatment of progressive MS remains a major challenge.

In 1972, when I was a young doctor studying neurology at the Peter Bent Brigham Hospital in Boston (now the Brigham and Women's Hospital), there were no treatments for MS. If I diagnosed a young man or woman with MS in 1972 and was asked, "What's the treatment, Doctor?" the only response was cortisone shots—hardly a treatment at all, given the disease's symptoms and progression. Today, when I sit with an MS patient with the relapsing form of MS and am asked that question, it takes me over thirty minutes to explain the many treatment options we have available. We have certainly come a long way, although we certainly have farther to go.

Earlier, I pointed out that we are just beginning to make the amount of progress with neurological conditions that we have made over past decades with other conditions—like the heart, for example. But there are distinctions even among neurological diseases. When I compare MS to other neurological diseases, I am struck by how little headway we have made against diseases such as Alzheimer's disease and ALS. It is hard to imagine that I have no more truly effective treatments to offer when I see a patient with one of these disorders today than I had forty years ago. Why did we make such advances in MS and not in these other diseases? Can we apply the principles that have led to success in MS to these other diseases? One key is that we succeeded in identifying specific disease mechanisms in MS. Looking back, our success in understanding the disease stemmed both from advances in our understanding of the immune system and in developing an animal model of MS that could be used to devise and test potential treatments. Another major reason for the advances in MS relates to MRI imaging. The MRI has allowed us to see the disease in the brain and to objectively monitor its progression and response to treatment—a huge development. Imagine treating heart disease without an electrocardiogram or tuberculosis without a chest X-ray—it would be very difficult indeed.

Fifteen years ago, I wrote a book entitled *Curing MS*. I chose that title because, frankly, that's what every disease is about: finding a cure. I was going to name that book *Taming the Monster*, the title of one of the chapters, and indeed MS is a monster we are trying to master. But I elected to use the word *cure* in the title and put it in the active tense. If I were to write a book about MS now, I would call it *MS: A Cure Within Reach*, because so much progress has been made in the past fifteen years.

In my book *Curing MS*, I tell the story of my childhood friend Normie, who developed MS in his twenties and became significantly disabled. We called each other by our childhood nicknames, Big Skinny and Little Fats, and many who have read *Curing MS* ask how Little Fats is doing. I visited him regularly on my trips to Denver, where my sister, son, and grandchildren live. He was in a rest home and at the age of seventy-three was dependent on a motorized wheelchair. He developed the progressive form of MS, a type of MS for which we have few effective treatments. Sadly, he passed away at the age of seventy-four.

The progressive form of MS, like Alzheimer's and ALS, is an escalating disease with few treatments. Although treatment came too late for him, Normie never lost his spirit, sense of humor, and love of life. When I last saw him, he said to me, "It's getting tougher, Big Skinny, you've got to do something about this crazy disease." The urgency I heard in his voice was not only for himself but also for his daughter, who has also contracted MS and has the treatable, relapsing form of the disease. I am helping to manage her case so she doesn't develop the progressive form, like her father.

I believe we are on the precipice of major advances in brain diseases, and there is a fascinating story to be told about that journey. Fighting against neurological diseases is the story of advances in basic science and technology and how they are applied to these diseases. It is the story of the scientists and doctors who both collaborate and compete. It involves big business and the billions of dollars that are to be made from new treatments. Most importantly, it is the story of those who suffer from neurological diseases and the suffering they must endure until a cure is found.

In this book, it is my intent to provide a window into how medical research is carried out in the twenty-first century. Scientists no longer can work alone, although an unexpected discovery by a single scientist can have a major impact. Our Center for Neurologic Diseases, over three hundred people strong, is now named in honor of Ann Romney, who, for over a decade, has been in my care in her fight against MS.

Disease is nonpartisan. Joining the Ann Romney Center board are members of the Clinton and Kennedy families. In my discussions with Ann and Mitt, I learned that messaging is important in political campaigns and that staying on message is crucial. Could this concept be applied to science and the goals of the Ann Romney Center? The answer is yes. We need to know where we are heading, and we need to do something bold, something transformational. To reflect our initiatives, I came up with three slogans to serve as principles for our research: Drilling for Oil, Breaking Down Silos, and Shots on Goal. Drilling for Oil refers to performing risky experiments, taking chances, and not being afraid to drill a dry well. We need to raise money to drill for oil. Breaking Down Silos refers to promoting cross-fertilization among disciplines and scientists. And Shots on Goal refers to the reason we do our research, to find new treatments. A shot on goal is a clinical trial of a new therapy. If you don't try to treat, you will never help, and you will never score.

In this book, we will journey to the laboratories where research is being carried out, both at the Ann Romney Center at the Brigham and Women's Hospital and labs throughout the world, where we will join the front lines in the search to understand and cure five of the most challenging diseases of the brain: MS, Alzheimer's disease, ALS, Parkinson's disease, and brain tumors. We will try to answer the frustrating question so many patients ask: What is needed to identify the perpetrators and stop the siege on their brains? Why is there no cure for these diseases?

CHAPTER TWO
Multiple Sclerosis

In the month after he entered medical school, Bob Engle noticed tingling in both feet after jogging. He didn't pay much attention, as it lasted for only a week and then disappeared. Later that year, he felt an overwhelming sense of fatigue, but he attributed it to his heavy workload and final exams. The feeling of fatigue also passed. Then, in his second year of medical school, he noticed that sometimes when he bent his neck down, he had the same tingling in his legs that he had after jogging. This, too, passed, and again, he didn't pay much attention to it. But in his third year of medical school, something happened that he couldn't ignore. He lost sight in his right eye. While watching a procedure in the operating room, he noticed a disturbance in vision, which he initially attributed to the bright lights in the room. But his vision became worse over the next forty-eight hours, and when he covered his left eye, he could only see a blur with his right eye. He immediately called his father, a cardiologist who worked in our hospital. They met at the ophthalmologist's office the next morning. The ophthalmologist found that the vision in Bob's right eye was 20/400, and when he examined the eye, he found that the optic nerve in Bob's right eye was swollen. He diagnosed optic neuritis, inflammation of the optic nerve. Knowing that optic neuritis was frequently associated with MS, he immediately ordered an MRI scan of Bob's head. The MRI of the brain was done three hours later and showed white spots consistent with multiple sclerosis. One hour later, I received a call from his father. Could I see his son? "When?," I asked. "How about this afternoon?" his father asked anxiously.

Given my workload, it is difficult for me to take on new patients, and even then, there can be a long waiting time for an appointment. In this case, I kept the clinic open until 5:30 PM and saw Bob Engle and his father that same afternoon. After taking Bob's history, performing a neurological exam, and reviewing the MRI, the diagnosis was clear.

"Your son has multiple sclerosis," I said to the father.

"Are you sure?" he asked.

"I'm sure."

"Are there treatments?" Bob asked me.

"Yes," I said, "and I think your prognosis is excellent."

"What about a cure?" the father asked. "Is there a cure for MS?"

Although this is a simple and very direct question, when one examines it in detail, it is more complicated than it appears. It is a question that begets another question, and I respond to the cure question posed by almost all of my patients with another question: "What do you mean by a cure?"

Listening to their responses, it has become clear that there are three definitions of a cure. In the first, a person who develops MS is placed on treatment and lives a full and normal life with no disability. In the second, a person with MS who has developed a disability is treated and the disability is reversed. In the third, a vaccine is given and, like polio, no one develops MS—MS is prevented. I first described these three types of cures in my book *Curing MS*. The National Multiple Sclerosis Society has adopted a version of them and now classifies the research they support as being in three categories: STOP disease progression, RESTORE what's been lost, and END MS forever. These three cures also apply to the other neurological diseases presented in this book, and we will explore the pathways to achieve these goals. Of the five, only MS is now positioned to achieve one of the cures (stop attacks of disease and prevent disability), although the path is clear for all the diseases to reach this benchmark.

I recently saw a man in our MS Center whom I have been following for fifteen years. Like me, he has two sons. When his MS was first diagnosed, I immediately treated him with a recently approved MS injectable medication. Later, he was switched to a newly developed oral medication, which he takes daily. The last time I saw him, he said, "You know, Dr. Weiner, I feel great, I don't even feel like I have MS." After I performed a quick neurological exam and checked that the MRI done the week before was stable, we spent the remainder of the clinic visit talking about our sons.

Unfortunately, not all cases are like his. There are patients who have attacks despite treatment and others who enter the progressive phase of MS for which we do not have effective treatment. Thus, there is the joy I feel when I can treat someone and I give them an A+ at the end of their visit, and the pain I feel when, no matter what I do, I am unable to help. All patients love getting an A+ no matter their age or station in life.

CURING DISEASE

To cure a disease, one must know the cause. This knowledge is how we cured polio. The path to a cure for polio began when it was recognized as a distinct clinical entity by Jakob Heine in 1840, and then when polio was identified as an infectious disease that could be transmitted from one person to another. The next step toward a cure was to identify the infectious agent. The poliovirus was identified in 1908 by Karl Landsteiner. However, once the virus was identified, a major hurdle remained. To make a vaccine, it was necessary to find a method to grow the virus in the laboratory. It was a thorny problem that was ultimately solved by John Enders and Thomas Weller, who were later awarded the Nobel Prize for their discovery. Although Jonas Salk's name is linked in the public eye to the polio vaccine and the cure of polio, he did not share in the Nobel Prize.

After initial studies, a national trial was begun in April 1954 in which 440,000 children between the ages of six and nine received the vaccine and 210,000 children served as controls and received a placebo injection. The dramatic results were announced a year later on April 12, 1955. Few of the children who were given the vaccine developed polio, while the unvaccinated children did. If we'd had the wrong virus and vaccinated with smallpox, for example, it wouldn't have worked. We needed to know the cause in order to cure polio. Jonas Salk's interest in paralytic diseases led him to another experiment in which he tried to develop a vaccine for MS. He based his MS vaccine on the theory that MS was caused by an abnormal immune response against a brain protein called myelin basic protein, or MBP. Salk vaccinated MS patients with MBP isolated from pigs, hoping that by doing so, he could cure MS or at least slow it down. Unlike his polio experiment, his MS experiment failed. He did not have the right cause.

What, then, causes MS? When I teach MS to medical students, I use a didactic approach in which I begin with the following question: What information would you need to understand the cause of MS? The first step of course is to recognize the clinical symptoms of the disease. The symptoms of MS were formally described in 1878 at the Salpêtrière in Paris by Jean-Martin Charcot. He codified what has come to be known as Charcot's triad:

1. *Nystagmus,* or jerky eye movements that can lead to double vision;
2. *Intention tremor,* or tremor that appears when one reaches for something; and
3. *Dysarthria,* or slurring of speech due to damage to fibers in the brain stem.

If you had the clinical symptoms of Charcot's triad, imagine trying to explain to a policeman that you are not drunk but have MS. Luckily, you can take a breathalyzer test.

Although these are classic features of MS, in the forty years that I have been treating patients with MS, the MS symptoms that immediately come to mind are similar to those that Bob Engle experienced—numbness or tingling in an extremity or across the chest or belly and blindness or loss of vision in an eye. I think this is because MS may be changing, and with MRI imaging, we are able to recognize it earlier and earlier. The other classic feature of MS is that in early stages of the disease patients usually recover from their attacks, even without treatment. There can also be silent MS. There are instances of MS discovered at autopsy in someone who had no symptoms during their lives. What makes MS difficult for both patients and their physicians is that MS is a protean disease that can affect people in many different ways.

THE CRIME SCENE

When I ask medical students what is needed to understand the cause of MS after recognizing the clinical symptoms, I use the analogy of a murder. What does the crime scene look like? In MS, the crime scene can be found in the brain and spinal cord. The heart, the kidneys, the liver, and the lungs are all normal. But when you look at the brain and spinal cord, something is clearly wrong. The brain consists of white and grey matter, and when you hold a normal brain in your hand, the grey and white matter are clearly demarcated and easy to see. But when you look at the white matter of the brain of an MS patient, with the naked eye, one can see blotches, areas that are no longer glistening white but have turned grey. Furthermore, instead of feeling soft to the touch, the tissue is hard, sclerotic. There are multiple areas of scar tissue scattered throughout the brain. These multiple scars give the name *multiple sclerosis.* There are also scars in the spinal cord, as the spinal cord is directly connected to the brain and serves as a conduit for electrical impulses traveling to and from the brain. But the nerves that exit the spinal cord are normal, with no blotches or sclerotic areas. So, the first report from the crime scene indicates that MS affects only the brain and spinal cord (the central nervous system) and has sclerotic areas that are not seen in other diseases of the central nervous system, such as Alzheimer's disease, Parkinson's disease, or ALS.

MULTIPLE SCLEROSIS

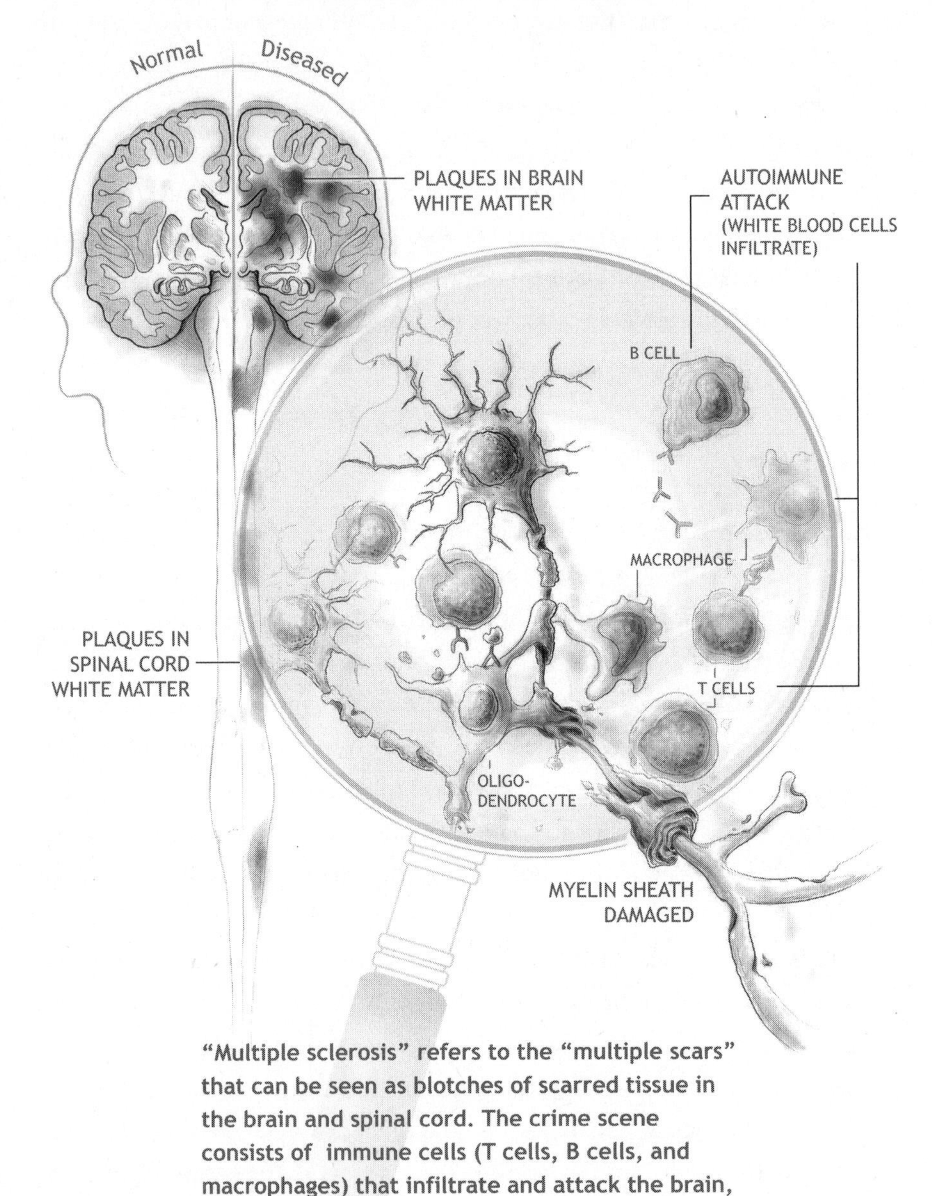

"Multiple sclerosis" refers to the "multiple scars" that can be seen as blotches of scarred tissue in the brain and spinal cord. The crime scene consists of immune cells (T cells, B cells, and macrophages) that infiltrate and attack the brain, causing damage to the myelin sheath.

The next step in examining the crime scene is to take a piece of the brain that has splotches and look at it under the microscope. The first microscopic description of MS was made in 1850, by the pathologist Santiago Ramon y Cajal, and a landmark paper was published by the Scottish pathologist James Walker Dawson in 1916. Microscopes at that time were primitive compared to what we have today, but they were strong enough to provide a major insight into the MS crime scene. In the brain and spinal cord of people with MS, there are cells that shouldn't be there—white blood cells. White blood cells are typically found in the bloodstream, so why were they in the brain and spinal cord?

In fact, it was not unusual for them to be there, as white blood cells are always leaving the bloodstream and entering the brain. White blood cells are part of the immune system, and the immune system evolved to fight infection. They are sentinels that examine the brain, make sure nothing is wrong, and then leave. Thus, a simple explanation of the crime scene is that there is an infection in the brain and the white blood cells entered the brain to fight the infection. Thus, curing MS should be easy. Find the infectious agent, vaccinate against it, and cure MS.

However, all attempts to isolate a virus or other infectious agent from the brains of MS patients have failed. There have been many false culprits identified, including the spirochetes that cause syphilis and viruses related to the AIDS virus. For many years MS patients were not allowed to donate blood for fear MS might be an infectious disease. This restriction has been lifted.

If there is no infection in the brain, why do the white blood cells go there and cause damage; and why don't they cause damage anywhere else in the body? The answers relate to what we now believe is the underlying cause of MS. MS is an autoimmune disease (immunity against oneself). The white blood cells are actually attacking the brain and spinal cord.

If a kidney or a heart is transplanted between people who are not identical twins, the immune system attacks and rejects the transplanted organ. It recognizes the transplanted kidney as foreign, like an alarm system warning against an intruder. The same type of white blood cells that we find in the brains of MS patients accumulate in the kidney after someone gets a kidney transplant if anti-rejection drugs are not given.

Although there are several theories about why the immune system attacks the brain as if it were a foreign object like a transplanted kidney, the most common idea is in fact linked to viruses or infectious agents and is termed molecular mimicry. The molecular mimicry theory states that there is an infectious agent such as a bacteria or virus that has a structure on it that is identical or very similar to a structure in the brain. When a person is infected by the virus the immune system attacks the virus, but when the white cells enter the brain, because a structure in the brain is identical to a structure on the virus, the immune cells attack the brain, thinking they are attacking the virus.

There are precedents for molecular mimicry in medicine. A prime example is what happens when strep throat leads to rheumatic heart disease. In rheumatic heart disease, the mitral valve of the heart is damaged a few weeks after infection with strep, because the *Streptococcus* bacteria and the mitral valve of the heart have identical structures. Thus, the immune cells attack the mitral valve believing they are attacking the *Streptococcus.* Treating a strep throat with penicillin or other antibiotics is important because eradicating the bacteria with antibiotics precludes the immune system from fighting the bacteria and mistakenly attacking the mitral valve of the heart.

Untreated strep throat can also lead to a disease of the brain called Sydenham's chorea, as there are structures in the brain that are similar to *Streptococcus.*

In fact, many viruses have structures similar to parts of the brain, and many people can be exposed to those viruses, but only some people's immune systems will attack their brains and cause diseases like multiple sclerosis. Guillain-Barré syndrome—a paralytic disease caused by an autoimmune attack against the myelin in the peripheral nerves—is often triggered by a preceding infection, by a variety of viruses or bacteria, including

Campylobacter. In Guillain-Barré, the nerves are able to regenerate, something that doesn't happen nearly as well in MS.

The virus most commonly associated with MS is the Epstein-Barr virus (EBV), which causes infectious mononucleosis. Virtually everyone has been infected by EBV, even if they have never had mono.

But if we've all been infected, why do only some people get MS? MS is a multifactorial disease, meaning that it's much more complicated than a virus triggering an attack against the brain. MS occurs in a genetically predisposed individual who encounters a constellation of environmental factors at the appropriate time in their life, including sunlight, vitamin D, smoking, and even the bacteria in the gut. Understanding how these factors contribute to the disease will be required to develop a cure for MS analogous to the cure for polio—in other words, a vaccine.

IMMUNE ORIGINS

Although MS is triggered by multiple factors that must align in a specific formation at a unique point in time, we believe that this alignment triggers one cell that initiates the process: a T cell, so named because it comes from the thymus gland in the neck. After the T cell matures in the thymus, it enters the circulation and orchestrates the immune response. There are two biologic systems that have memory; one, of course, is the brain, and the other is the immune system. If you've been infected with polio or smallpox, your immune system remembers it, and you are protected from having the infection again.

As discussed, we believe the immune system attacks the brain in MS. One possible explanation for why some people get MS and others don't is that in patients with MS, T cells generated in the thymus react with the brain, whereas people who don't get MS don't have T cells that can react with the brain. Although this sounds like a plausible explanation, it doesn't work, as we all have T cells in our bloodstream that react with the brain and can enter it and cause MS.

The reason some people's brain-reactive T cells enter the brain and cause damage while other people's do not is that there are different subtypes of T cells that react with the brain. Some can cause disease, and some don't. One of the major advances in immunology has been the identification of different T cell subsets. When I was first learning immunology in 1975, I trained with Henry Claman at the University of Colorado. He made the basic discovery that there are two major types of white blood cells in the immune system. He classified them as T cells and B cells. B cells are responsible for making antibodies, and they need help from T cells to do so. In the following decade, T cells were subdivided into subsets called Th1 and Th2 cells, based on the different chemicals they produce. In 1994, we described a third type of T cell, which secretes a chemical called TGF-beta and which we named a Th3 cell. Since then, scientists have identified a vast number of T cell subtypes, based on markers on their surface, the chemicals they secrete, and the profile of the genes inside them.

All in all, there are over two hundred markers on the surface of T cells that can be used to sort and classify them; additionally, there are thousands of genes inside the cell. We hang large posters in the laboratory that pictorially represent the complexity of the immune system and interactions between all the different cells. The immune system has often been described in military terms (one subset of T cells is called "killer T cells"). With the many arrows and pathways depicted, our posters resemble a complex battlefield map from World War II. The success of our developing treatments for MS has been directly related to our understanding of the intricate workings of the immune system.

We prove that we understand what causes a disease, as well as the mechanisms involved, when we interrupt or modulate the disease mechanism and the disease process is ameliorated. Thus, if MS is caused by T cells entering the brain, treatments that interrupt or stop their entry should help ameliorate the disease.

As we discussed, T cells are sentinels that patrol the body looking for enemies to attack; when they enter the brain, they mistakenly attack healthy tissue. A T cell enters the brain by attaching to a specific receptor on the wall of a blood vessel and exiting the bloodstream. The attachment works through a lock-and-key mechanism involving a structure on the T cell and a structure on the blood vessel. One of the most effective drugs for preventing MS attacks, called Tysabri (natalizumab), binds to the structure on the T cell and prevents it from attaching to the blood vessel. In this way, killer T cells are trapped in the blood and can't get into the brain. Tysabri is one of the most effective MS drugs in reducing attacks such as the tingling or loss of vision that Bob Engle experienced, proof positive of the importance of cells moving out of the bloodstream into the brain and the role of the immune system in MS.

Unfortunately, blocking the movement of T cells from the bloodstream into the brain has undesired consequences. The immune system evolved to fight off infection, and T cells serve as sentinels that enter the brain to prevent unwanted infections from gaining hold there. When natalizumab blocks MS-inducing T cells from entering the brain, it also arrests other T cells with protective function. This has resulted in an increased incidence of a fatal brain infection called progressive multifocal leukoencephalopathy (PML) in people treated with natalizumab. When Tysabri was first approved in 2003 by the FDA, we treated hundreds of patients in the first months, until the drug was removed from the market because of PML infection. In subsequent years, it was carefully reintroduced to the market and given to people who underwent blood testing to assess their exposure to the virus that causes PML. This illustrates the double-edged sword of modifying the immune system to treat MS. One of the major challenges we face in treating MS is to control the immune system without causing untoward side effects.

There are currently over fifteen drugs approved by the FDA to treat MS. When I began my professional life, there were none. How did we get there, and what lessons can we learn from our success in MS that can be applied to diseases such as Alzheimer's and ALS—conditions for which we have no truly effective treatments?

Drug development for MS or for any disease is not always as logical as the story for Tysabri. For example, the first drug to be approved by the FDA in 1993 for the treatment of MS, Betaseron (beta interferon), was tested in MS patients for its antiviral properties on the theory that a virus was causing MS. We now know that beta interferon helps MS because of its effect on the immune system, not its antiviral properties.

CYCLOPHOSPHAMIDE AND PLASMA EXCHANGE

In 1976, when I took my first faculty position at the Brigham and Women's Hospital and Harvard Medical School, there were two major theories about the cause of MS:

1. MS is caused by an infectious agent, such as a virus,

   or

2. MS is an autoimmune disease caused by a misdirected immune system.

Because I was so interested in understanding and finding a cure for MS, I embarked on training in both virology and immunology. At that time, there was evidence that the measles virus might be associated with MS, and a paper was published in the *New England Journal of Medicine* demonstrating that MS patients had abnormal reactions to measles virus. I was working in the lab of virologist Bernie Fields at the Harvard Medical School when the National MS Society solicited a grant from him to study measles. We were never able to repeat the measles findings that were published in the *New England Journal of Medicine.*

My first independent grant to study MS was related to immune cells called natural killer cells and how they may be defective in fighting viruses. The grant was given by the Kroc Foundation, established by the Kroc family of McDonald's fame. As it turned out, their interest stemmed from the fact that members of the Kroc family suffered from autoimmune diseases including MS, rheumatoid arthritis, and diabetes. I was later awarded an endowed chair at Harvard from the Kroc family named after Robert L. Kroc, the brother of Ray Kroc, who founded McDonald's.

I had far more success pursuing the immune approach than I'd had with the viral approach. At that time, Steve Hauser was a neurology fellow at Massachusetts General Hospital and was interested in MS. He came to talk to me at Bernie Fields's lab about possible approaches to treating MS. We both knew that if MS is an autoimmune disease, one should be able to slow the progression of the disease by damping or suppressing the immune system.

The most classic autoimmune diseases are linked to autoantibodies, antibodies that react to a person's own tissue, even though the tissue is normal. Myasthenia gravis is one such example. People with myasthenia have weakness in their eye muscles and other muscle groups because their bodies produce antibodies that block transmission of signals at the junction of nerves and muscle. The definitive proof of this link was an experiment in which antibodies were removed from the blood of patients with myasthenia gravis and injected into mice. The animals developed muscle weakness, just like the patients. This led to treating myasthenia patients with plasma exchange in which patients were hooked up to a machine and their blood cleansed of the disease-causing autoantibodies.

There was some evidence suggesting that MS patients might have autoantibodies, although it was certainly not as clear as it was for myasthenia. And no one had ever transferred MS to a mouse by injecting antibodies from MS patients. I have always been therapeutically aggressive, and even if no one had identified autoantibodies in MS, it didn't mean they weren't there. Why not treat MS patients with plasma exchange and go directly for the answer?

The big questions were whether it was safe and how difficult it would be for patients to undergo the treatment. Thus, I underwent two plasma exchanges myself, one in which the plasma was removed and the other in which I received a sham exchange in which blood was put through the machine but then given back to me. I couldn't tell the difference between the two procedures and I suffered no ill effects, but I didn't tell my wife until it was over. After that, along with Dave Dawson at the Brigham, we treated a small number of MS patients with plasma exchange. The patients seemed to do better, although the improvement didn't last.

Another way to modulate the immune system is to use drugs that are general suppressants of the immune system. Remember, in kidney transplantation the immune system attacks the kidney because it is foreign to the recipient, so suppressing the immune system has become a mainstay of kidney and heart transplantation. In MS, the immune system attacks the brain and spinal cord because it mistakenly thinks they are foreign. It follows that drugs that suppress the immune system might work in treating MS.

At Massachusetts General Hospital, Steve Hauser was taking care of an MS patient with breast cancer, who'd received cyclophosphamide (an immune suppressant) as part of the chemotherapy for her breast cancer. Her MS got dramatically better. There had been other positive reports of treating MS with drugs that suppress the immune system, but never in a controlled trial. Steve Hauser and I decided to assume that MS is indeed an autoimmune disease and to conduct a trial in which we compared the two different approaches to treating MS: plasma exchange and general immunosuppression using cyclophosphamide.

One of the biggest challenges in performing clinical trials is to perform a placebo-controlled trial in which the patient doesn't know which treatment they are receiving. There are striking examples of positive clinical results that cannot be replicated in a controlled trial. One such occurrence was in Parkinson's disease, in which symptoms of the disease appeared to be helped by implantation of cells into the brain; a subsequent trial of a sham operation, however, showed the same result.

Sometimes, however, it is not possible to perform a placebo-controlled trial. This was the case in the trial Steve Hauser and I designed. We decided to treat three groups: the first received a steroid-stimulating drug called ACTH, which had shown some positive effects in MS; the second group received ACTH plus high-dose cyclophosphamide; and the third group received ACTH plus plasma exchange and low doses of cyclophosphamide. Because patients lose their hair with chemotherapy and are hooked up to a machine for plasma exchange, all patients knew which treatment they were receiving, although they were randomly assigned to one of the three groups when entering the trial.

It was a bold experiment, and I remember examining patients at the clinical research center at the Brigham, carefully administering cyclophosphamide until their white blood cell count began to drop. MRI imaging did not exist at the time we performed our trial, so we could not see what was happening in the brain; we only had clinical measures to calculate results. Today, MRI imaging of the brain is an integral part of every clinical trial, and every FDA drug approved for MS has shown positive effects on the MRI.

Our results were dramatic: 80 percent of people treated with the high-dose intravenous cyclophosphamide plus ACTH stabilized or improved; 40 percent who received the plasma exchange plus ACTH plus low-dose oral cyclophosphamide benefitted; but only 20 percent who received the ACTH were helped. Some patients significantly improved. Our results were published in 1983 in the *New England Journal of Medicine*, one of the most prestigious clinical journals in the world. It created quite a stir—it was the first time in a controlled trial that a drug was shown to affect MS.

The *New England Journal* is published on Thursday. Each week's issue is under embargo until 5 PM Eastern Time on the Wednesday night prior to the Thursday publication date. Media subscribers receive access the previous Friday at 10 AM ET, after which time authors may speak to reporters. This schedule is beneficial to the *Journal*, but it means the public will hear the results before doctors and scientists can read and evaluate the article. Thus, I often receive calls from patients asking about medical news before I am able to read the results for myself.

I learned about neurology and science in medical school and on the hospital wards. I learned about medicine and the media on the job. The hospital realized the trial would be a big story and expected so much interest that they organized a press conference. There was tension among the Brigham and Women's Hospital, where I was based, the Children's Hospital, where plasma exchange was performed, and Massachusetts General Hospital, where Steve Hauser was based, over where the press conference would be held. In the end, it was held at Brigham with a doctor from each hospital present. Today, all the top journals, whether scientifically based, such as *Nature* and *Science*, or clinically based, such as the *New England Journal* and the *Lancet*, promote their journals by encouraging press coverage of the articles they publish. The top medical institutions market themselves with slick magazines and try to get their names in the news. The public is fed with lists of the top hospitals and top doctors in the country. It is a necessary form of validation, but at times it's disconcerting. There was a picture of me in the *New York Times* and a local news program was broadcast from my laboratories. Because cyclophosphamide is a drug used for chemotherapy, one of the reporters asked me, "Dr. Weiner, what would you rather have, cancer or MS?" So much for the news media and their level of scientific inquiry as they balance scientific reporting with sensationalism.

Our article created a great stir, and physicians began using cyclophosphamide in severe cases of MS. Because it was an approved drug for other conditions, it didn't need FDA approval for use in MS, and doctors had the ability to prescribe it, although a patient in California successfully sued her insurance company when it wouldn't cover treatment with the drug. However, we soon discovered that although an intensive short course of treatment with cyclophosphamide stopped MS progression and even reversed disability in some patients, eighteen months later, symptoms began progressing in patients again. These studies were the first to show that one had to continually modulate the immune system so that it wouldn't become activated and attack again. Indeed, today, MS therapy is usually given on a consistent basis without stopping.

Following the study, we set up the Northeast Cooperative Treatment Group, in which all MS patients were given cyclophosphamide, as we did in our *New England Journal* publication, and then randomized to receive monthly outpatient infusions of cyclophosphamide, as was used in another autoimmune disease called lupus erythematous. We found that MS patients who received the monthly outpatient infusions maintained their stability longer.

A major twist in the cyclophosphamide story occurred when a Canadian group led by John Noseworthy and George Ebers attempted to replicate our findings. Biological truth in science and medicine is measured by the ability to independently replicate crucial findings. We always do it in the lab to make sure there are no artifacts in our results. For example, we once found that a treatment that appeared to help mice with the animal model of MS was not related to the treatment but to a contaminant in the test tube where the treatment was prepared. We discovered the artifact when another lab in our center couldn't repeat our results. Sometimes the failure to replicate is not due to artifact but to outright scientific fraud, as was the case of a Japanese scientist who falsely reported in the prestigious journal *Nature* that she had discovered a simple new method for creating stem cells. Because her results were fraudulent, they were impossible to replicate.

If cyclophosphamide really helped MS as we demonstrated in our randomized trial, and as European investigators Otto Hommes and Richard Gonsette had shown in their open-label studies (studies in which both the doctor and the patient knew what treatment they were being given), then the results should replicate. Investigators in Canada designed a study to replicate our findings and published their results in 1991, eight years after our *New England Journal of Medicine* article. Unexpectedly, the Canadian study found no difference between cyclophosphamide and plasma exchange versus placebo, also in a randomized trial. They performed more rigorous blinding by having placebo patients receive sham plasma exchange and take placebo pills. The results were presented at the annual American Academy of Neurology meeting in Miami the year before they were to be published. I was chairman of the scientific session in which the results were presented, and I remember sitting on the dais holding my breath as the data were shown. The results were negative. Even worse, the Canadian team criticized our work for not having a proper control group and accused us of faulty science. By this time, doctors from around the country were treating MS patients with our published cyclophosphamide protocol and telling us of dramatic results. We had moved on to the next scientific question and were treating 256 patients with cyclophosphamide induction followed by monthly outpatient cyclophosphamide infusions as part of the Northeast Cooperative Treatment Group, a study that we published in 1993. As Steve and I pored over the Canadian study, we scratched our heads to find out why there were differences between their study and ours. There was no question in our minds that we had seen dramatic effects in our patients treated with cyclophosphamide. There was also no question that the Canadians had performed a very careful study.

As it turned out, both our study and the Canadian study were correct. The difference related to the type of MS each group was treating and foreshadowed the central challenge facing the treatment of MS today: progressive MS. The title of our *New England Journal of Medicine* publication was "Intensive Immunosuppression in Progressive MS." We defined *progression* as a decrease in one or more points on a clinical disability scale, consisting of either a continuous decline or a continuous decline in association with exacerbations. We specifically avoided patients with purely relapsing-remitting disease, as relapsing disease is less predictable and can spontaneously improve. The Canadian study was published in the *Lancet* with the following title: "The Canadian Cooperative Trial of Cyclophosphamide and Plasma Exchange in Progressive Multiple Sclerosis." However, there appears to have been a big difference in the type of progressive patients that were treated in each study. In retrospect, we were treating patients who were not purely progressive but had a mixture of relapsing and progressive disease, whereas the Canadians treated patients who for the most part were purely progressive without a relapsing component. Thirty years later, the FDA would classify progressive patients who had concurrent relapses as "active progressive," whereas those without concurrent relapses were classified as "non-active" progressive.

A major deficiency in both of our studies was that MRI scans were not then available, and our patients were chosen on purely clinical grounds. No modern study in MS is now done without MRI imaging. Betaseron was approved for MS because MRIs showed the treatment was shown to reduce MS activity in the brain. Although the FDA then approved the drug Copaxone (glatiramer acetate) based solely on clinical data, it wasn't until an effect was observed in a subsequent MRI study that Copaxone was fully accepted by the neurological community.

The first study of MRI in patients treated with cyclophosphamide was published in 1999 and was performed by Henry McFarland, chief of the Neuroimmunology Branch at the National Institutes of Health (NIH) in Bethesda. The article was entitled "Reduction of Disease Activity and Disability with High-Dose Cyclophosphamide in Patients with Aggressive Multiple Sclerosis." Five patients who were experiencing rapid clinical deterioration despite treatment with other modalities were placed on monthly infusions of cyclophosphamide and underwent monthly MRI imaging that included injection of gadolinium (a contrast medium) into the bloodstream. The response was dramatic. Patients improved clinically, and there was no evidence of gadolinium in the brain, a phenomenon called gadolinium enhancement.

Gadolinium is a compound that is given intravenously at the time of MRI and lights up on the scan, showing active inflammation. In someone without MS, it circulates in the bloodstream and doesn't enter the brain because of the blood-brain barrier, so there is no gadolinium signal in the brain. In MS there is a breakdown in the blood-brain barrier, and the gadolinium leaks into the brain at the sites affected by MS. Because MS attacks are caused by the movement of white blood cells from the bloodstream into the brain, when these cells enter the brain, they open up the blood-brain barrier and gadolinium can enter the brain as well. David Miller in

England demonstrated this principle in a classic MRI study in which the drug Tysabri was given intravenously. Tysabri blocks a receptor on white blood cells that is needed for the cells to leave the bloodstream and enter the brain. Once a month, Miller performed MRI imaging on patients who were being treated with Tysabri and found, beginning one month after starting therapy, the virtual disappearance of gadolinium-enhancing areas in the brain.

Although it was a trying time for me as the controversy surrounding the effect of cyclophosphamide in MS played itself out over two decades, a series of studies performed by investigators around the world helped clarify the role of cyclophosphamide in MS and set the stage for testing the role of strong suppression of the immune response in the treatment of MS.

These studies brought into focus the difference between relapsing and progressive forms of MS and how a different category of drugs would be needed to treat the two forms. In the two decades following our 1983 *New England Journal of Medicine* publication, both our group at Brigham and other MS specialists primarily used cyclophosphamide given as once-a-month outpatient infusions, just as we had developed in our Northeast Cooperative Treatment group protocol. This monthly approach allowed for easier and prolonged administration that didn't require hospitalization and wasn't associated with hair loss. In addition to McFarland's MRI study of cyclophosphamide boosters, six other investigators in the United States and Europe showed a positive effect of cyclophosphamide infusions given to patients with rapidly progressive MS. We reviewed our experience with cyclophosphamide boosters and found that response to therapy was linked to the duration of progressive disease; the shorter the duration of progression, the better the response. This is because there is still a relapsing component of MS in the early stages of progressive disease.

This was one of our first clues that the disease process in MS changes when a patient transitions from the relapsing to the progressive phase. Without treatment during the relapsing phase, over 70 percent of patients eventually become progressive. In relapsing MS, patients have discrete attacks and recover. The relapses are caused by white blood cells that move from the bloodstream into the brain.

In the progressive forms of MS, patients have slow progression of disease without relapses. Two things happen. First, the inflammation gets trapped in the brain, and second, the degenerative processes that get triggered in the brain are independent of inflammation. The disease process is no longer dependent on cells moving from the bloodstream into the brain. This is why there is not a great deal of gadolinium enhancement in progressive forms of MS—because white blood cells are not disrupting the blood-brain barrier and entering the brain.

Betaseron, the first drug approved by the FDA for MS, was approved for relapsing MS based on its positive effects in reducing the number of MS attacks and a pronounced effect on decreasing MS lesions as shown on MRI. Following its approval for relapsing MS, it was tested in a European trial for progressive MS and showed positive results on slowing disability as measured by gait and coordination. However, a North American trial of Betaseron in progressive MS was negative. In fact, the study was stopped early because an interim analysis showed no effect on progression. Henry McFarland, who had performed the MRI studies in patients treated with cyclophosphamide, was chairperson of the independent monitoring committee for both Betaseron studies. His explanation of why the European and North American studies differed was simple: Patients in the European trial were still in an earlier relapsing phase of the disease; there was an inflammatory component that contributed to their progression. This was not the case for patients in the North American trial, who had entered a later stage of the disease, or non-active progressive MS. Of note, other interferon preparations also did not show a benefit in progressive MS.

In addition to Steve Hauser, a neurologist named Bob Brown was a fellow in my lab in 1985 and was interested in understanding and treating ALS. Bob is now one of the world's leaders in the study of ALS and chairman of neurology at the University of Massachusetts. Because of the possibility that ALS might in some way be related to the immune system, we admitted ALS patients to the hospital and treated them with the same cyclophosphamide regimen Steve Hauser and I used in MS, although the crime scene in ALS did not have the inflammation seen in MS. Unfortunately, we saw no benefit.

Cyclophosphamide, a chemotherapy drug, was never approved for the treatment of MS by the FDA. Approval of a drug by the FDA is a complicated and expensive process that is virtually always championed by the pharmaceutical industry. There was no corporate sponsor for cyclophosphamide because it was a generic drug not on patent. Another chemotherapy drug, mitoxantrone, was approved by the FDA for MS, though it is seldom used because of its heart toxicity and risk of causing leukemia. In 2005, doctors reported that a forty-eight-year-old Brazilian woman who had a severe relapsing form of MS was accidently given a large dose of cyclophosphamide, after which her disease went into remission for seven years. Investigators at Johns Hopkins have tried very high doses of cyclophosphamide in an attempt to induce a permanent remission in highly active forms of MS. Two other approaches that have attempted to treat MS by massively suppressing the immune system are treatment by a monoclonal antibody called Lemtrada and autologous stem cell transplantation, explained below.

Monoclonal antibodies were discovered at Cambridge University by Cesar Milstein in 1975, for which he was awarded the Nobel Prize in Medicine. Monoclonal antibodies have revolutionized medicine, both for treatment of disease and their universal use in virtually all aspects of medicine and biology. An antibody is a Y-shaped structure that binds to its target via the tip (V portion) of the Y. When we receive a polio vaccine, we make antibodies against the poliovirus that inactivate the virus and make us immune to polio. Those antibodies bind to polio but would not bind to another virus, such as the smallpox virus. This principle is one of several being used to treat COVID-19: make monoclonal antibodies against the coronavirus and inject them to inactivate the virus.

Milstein discovered a technique to immortalize, or clone, the cell making the antibody so that billions of identical cells could be grown and unlimited amounts of a uniquely specific antibody could be made. Because the antibodies come from a single clone, they are called monoclonal antibodies.

Lemtrada is a monoclonal antibody developed by Herman Waldmann at the Cambridge University pathology department under the name CAMPATH-1. In 1982 it was initially used for the treatment of leukemia. Lemtrada binds to a structure on immune cells (present on both T cells and B cells) and kills them, causing profound suppression of the immune system. The immune cells then slowly return to normal as the immune system reconstitutes itself.

In 1991, Alastair Compston and Alasdair Coles in Cambridge, England, tested CAMPATH-1 in MS by administering it to patients with progressive disease, but they found no effect. They then turned to patients with relapsing disease and found a dramatic effect on relapses, disability, and MRI changes. In some patients, a five-day course led to complete remission lasting for several years.

The pathway to FDA approval of CAMPATH-1 for MS was a tortuous one that took two decades. The drug is administered intravenously for five days and then again for three days one year later. However, there are major side effects with Lemtrada, including infusion reactions, infections, and the development of other autoimmune diseases, especially in the thyroid. Although Lemtrada is one of the more effective drugs we have for the treatment of relapsing MS, it is used only in select patients because of its side effects.

Borrowing from the field of oncology, as we did when we gave cyclophosphamide to MS patients, investigators have used hematopoietic stem cell transplants, a form of bone marrow transplantation, to treat selected patients with aggressive, highly active MS. Bone marrow transplantation was developed by E. Donnall Thomas to treat leukemia, for which he was awarded the Nobel Prize in Medicine in 1990. To rid the body of cancerous leukemic cells, powerful chemotherapy and radiation are given, and the bone marrow, which is continually generating cells of the immune system, is killed in the process. The bone marrow is harvested prior to chemotherapy and then reinfused.

In MS, it is the body's immune system that is the cancer. Could MS be cured by deliberately destroying the abnormal immune system and then restoring it? In autologous stem cell transplant (often used in the treatment of leukemia) stem cells are harvested from the patient's bone marrow, after which the patient is given large doses of immunosuppressive drugs, including cyclophosphamide, to wipe out the immune system. The patient's previously harvested stem cells are then reinfused, and when the immune system regenerates, it is no longer diseased and able to attack the brain. To date there are published reports of more than six hundred patients with MS who have been treated with bone marrow transplants. Some of the results have been dramatic, with recovery of neurological function and absence of disease activity both clinically and on MRI.

There are different levels of immune system ablation that have been given to MS patients prior to bone marrow transplantation, and it has been argued that the stronger the ablation, the better the result. As would be expected, such strong treatment is not without risk, and there have been deaths. Similar to what happened with cyclophosphamide and Lemtrada, the positive effects of bone marrow transplantation are primarily observed in those who are in an active inflammatory relapsing stage of the disease, not in those with noninflammatory progressive MS. Unfortunately, such strong therapy doesn't help where it may be needed most—in patients with progressive MS, those with advanced disease, and severe disability. Because the words *stem cell* are part of the name of the treatment regimen, patients often mistake this form of immunosuppression with the concept of stem cell therapy, in which one attempts to replace damaged brain cells with fresh brain cells grown from stem cells.

A problem that plagues researchers investigating treatments like bone marrow transplantation is finding an appropriate control group. Because there can be no true placebo group, there must be an active comparator, as we had with our trial of cyclophosphamide, in which we compared cyclophosphamide to ACTH and a plasma exchange regimen. A similar approach was taken with Lemtrada, which was compared to the high dose interferon, Rebif. Even so, the FDA did not initially approve Lemtrada in 2013 despite it performing far better than Rebif and its having been approved in Europe. The FDA ruled that Lemtrada hadn't been proven safe and effective and cited poor study design because the Phase 3 trials of the drug lacked a placebo control group and was not blinded in that both patients and researchers knew the treatment they were receiving. After lobbying by the National MS Society and patient advocate groups, Lemtrada was finally approved a year later, in 2014. Because of serious potential side effects, a special risk management strategy was put into place for patients receiving Lemtrada, including monthly blood tests to screen for thyroid and blood toxicity.

Although most centers don't offer bone marrow transplant to MS patients, there are two large consortia that are treating select patients with bone marrow transplantation. At our center, in the rare case of rapidly progressive disease that requires immediate treatment, we hospitalize patients and use large doses of cyclophosphamide. One of the advantages of cyclophosphamide is that it crosses the blood-brain barrier and enters the brain, where it can immediately kill inflammatory cells. Some cases are dramatic: a thirty-five-year-old man whose MS progressed so rapidly that he was intubated and put on a respirator despite treatment with plasma exchange received two courses of cyclophosphamide and recovered. I now see him in our clinic, where he walks normally. A medical student with severe inflammatory disease visible on brain biopsy was able to graduate medical school and begin her family after receiving cyclophosphamide. But not all stories have happy endings. When I greet patients in our MS Center, I am often confronted by cases like a severely disabled young woman whom we couldn't help despite high-dose chemotherapy with cyclophosphamide. I wonder if stronger treatment would have worked for her.

As we will discuss below, we now have highly effective treatments for relapsing forms of MS. This raises an important question. Should a cardinal principle used for the treatment of cancer be adopted for our MS patients—treat as early as possible and with as aggressive therapy as possible to stop the disease in its tracks?

MS isn't cancer, and not all people who develop MS are destined for a wheelchair, though initially we don't know who is and who isn't. When I first cared for people with MS, it was frustrating for me to diagnose MS and have the patient ask, "What's the treatment, Doctor?" I could only answer that there was no treatment, but I was certain that one day there would be. That frustration is one of the factors that led me to the laboratory. That frustration still exists when confronting a patient suffering from Alzheimer's disease or ALS.

Occasionally, I will have a patient who doesn't want treatment for MS. I share with them what I call the Boxer Analogy. I tell them that confronting MS is like being in a boxing match. You are in the boxing ring, and your opponent is MS. You have to fight many rounds against the MS boxer. Each round lasts a year. It is likely that you will be hit at least once a year, maybe more. Being hit is a physical MS attack, such as going blind in an eye or losing coordination. Being hit could also be a silent attack that you aren't aware of and can only be seen on MRI imaging. Although you may be knocked down, initially there are few knockout punches, and you will be able to stand and fight again. However, the more times you are hit and knocked down, the harder it is to stand up again until you are no longer able to walk. MS treatments put a shield between you and the MS boxer, so you are not hit at all or only experience mild impact. If the MS boxer is strong, like Muhammad Ali, it may be hard to protect you. If MS is a child, you may not need as much protection, although MS is seldom like a child.

After I explain the Boxer Analogy, patients sometimes ask whether I can get them out of the ring completely so there is no chance of getting hit. Then there would be no need for medicine. Unfortunately, we still can't do that. However, for the first time we are beginning to ask whether treatment can be stopped when the disease has been stable over a period of time or when a patient reaches a certain age. Of course, the ultimate cure is a world where no one ever has to get into the ring with the MS Boxer. That scenario would involve giving an MS vaccine to everyone at risk for MS.

Developing an MS vaccine would require an experiment analogous to the poliovirus experiment, though it would take longer than seven months to see if it was effective. One would have to vaccinate thousands of children with the MS vaccine, have an equal number that weren't vaccinated, and then follow both groups for twenty-five years to see if the vaccine prevents MS. One day, we will do this experiment. I'll explore strategies for developing an MS vaccine later in the book when we discuss the microbiome.

There is the occasional patient who decides, against my recommendation, not to take MS treatment. Most of them ultimately receive treatment when they have an attack or disease activity is found on MRI imaging, which is done every six to twelve months. I have the rare benign patient who does well for years without treatment. I evaluated some of these patients before the first FDA-approved drug for MS became available in 1993; they were doing well and decided not to go onto therapy. This raises the questions: Is there a benign form of MS? Are patients taking risks by being off treatment early in their disease? Initially, a person with benign MS was defined as a person who has the disease for ten to fifteen years and shows little if any disability. It was originally defined by Stanley Hawkins in Ireland in the era before treatment. As it turned out, as those people were followed, many did in fact develop disability. The question today is how to define benign MS in the current era of treatment. The rare patient who has MS but has never been treated and doesn't accumulate disability could be considered to truly have benign MS. A second category is someone who has been on therapy from the start of their disease and acquires no disability; in these cases, treatment tamed the disease into becoming benign. There are many such patients at our MS Center.

Once treatments for relapsing forms of MS were FDA approved in the 1990s and we learned that the MRI could identify patients whose disease was active as measured by the appearance of new spots on the MRI, a new era of MS began. With this era came new challenges and a novel series of clinical and scientific questions. The primary question was related to understanding how individual patients did with treatment over time. In the two decades before the turn of the millennium, the central question in MS was whether there was any treatment that could treat MS attacks and lessen disability. One of the major questions then became: How effective are the treatments? If a twenty-five-year-old was diagnosed with MS and immediately began treatment, what would they be like twenty years later? For diseases such as Alzheimer's disease and ALS, we are still in a pre-treatment era. We have limited disease modifying treatment, and the major question for these diseases is very basic: How can we make a treatment breakthrough? Does the work with MS provide clues for solving these other neurological diseases?

TRACKING MS: THE CLIMB STUDY

In 1980 the central issues concerning MS were finding a treatment that could affect the disease course and determining whether MS was caused by an overactive immune system or a latent viral infection in the brain. We addressed these issues with cyclophosphamide and plasma exchange studies. Twenty years later, with the availability of FDA-approved therapies for relapsing MS, a different question arose. As we began to treat patients, it became clear that MS was not a uniform disease and patients responded differently to therapy. The question shifted from whether or not we could find something that affected the course of the disease in the short term, to what happened to patients over time and what made one patient different from another?

In order to address this, we set up a longitudinal MS cohort study along the lines of the Framingham Heart study, a landmark study initiated in 1948 to identify common factors that contributed to heart disease by following a cohort of subjects over time. We needed an approach that was scientifically grounded and not simply based on a clinical exam. The major advance was using MRI imaging with which we could actually see the disease. This led us to establish one of the first integrated MS Centers at the Brigham and Women's Hospital, which combined MRI imaging, clinical exams, and collecting blood samples from all patients on a yearly basis. Because infusions were also now being given to MS patients, we needed an MRI machine, an infusion center, and a clinical facility.

I spoke to the radiology department and found that there wasn't capacity to perform routine MRIs on all our patients. Although we had received a major grant from the NIH a few years earlier to study MRI use in MS, we needed to buy an additional MRI machine. However, the hospital didn't have the money although they agreed to give the space, and the National Multiple Sclerosis Society couldn't invest in local infrastructure.

Central to solving neurological disease is deciding which scientific approach to take, which questions to ask. Equally important is obtaining the funds to explore the questions. Traditional sources of funding are often not sufficient, and philanthropy provides the needed financial boost. I believe that every academic research endeavor needs philanthropic support.

Dennis Selkoe and I had already been raising money from grateful patients to help bolster our research. Dennis is an expert in Alzheimer's disease and codirects the Ann Romney Center for Neurologic Diseases with me. I was able to raise $4 million to purchase an MRI and pay for the necessary renovations for the MS Center. Soon after that, we recruited Rob Bakshi to join Charles Guttmann and run our MRI program.

With the new MS Center in place, we initiated the CLIMB study in 2000. CLIMB stands for Comprehensive Longitudinal Investigation of Multiple Sclerosis at Brigham and Women's Hospital, and it's our human laboratory for the study of MS. Patients enrolled in the CLIMB study receive clinical evaluations every six months, yearly MRIs, and yearly blood sampling. All data are placed in a comprehensive database. Now headed by Tanuja Chitnis at our center, CLIMB has enrolled over two thousand patients to date and has been tracking some patients for over fifteen years. More than 150 scientific publications have come from the CLIMB study, ranging from topics such as how environmental factors affect MS, to the effects of different treatments on the course of MS, MRI subtypes of MS, and the investigation of blood biomarkers in MS. Of all the questions we are asking in CLIMB, one of the essential ones is to what degree are we curing MS. As previously discussed, there are three definitions of cure for MS: a vaccine to prevent MS, stopping all disease activity, and rebuilding a damaged nervous system.

At CLIMB we're looking at the degree to which we are stopping all disease activity, a concept called NEDA, or No Evidence of Disease Activity. There are three measures we use to assess MS, which are used by the FDA to approve MS drugs: relapses, disability progression, and changes on MRI. NEDA means that there are no relapses, no progression, and no new MRI lesions or abnormalities. The emergence of NEDA as a concept in evaluating disease developed from rheumatology. Today, clinical trials in rheumatoid arthritis, and therapy in the clinical practice of rheumatology, are aimed at disease-free status. In MS, clinical trials have also begun to incorporate disease-free status as an outcome measure.

We took advantage of the CLIMB study to determine how many patients had achieved disease-free status over a seven-year period. We found that 46 percent had NEDA at one year but only 7.9 percent had such status after seven years. We had not stratified patients according to treatment, and most patients hadn't been treated with some of the more effective drugs that had been recently introduced. Trials of more potent drugs such as Tysabri and Lemtrada had NEDA levels below 40 percent after two years. Autologous stem cell transplant, the strongest anti-inflammatory treatment we have, has the largest NEDA at three years—75 percent. However, this falls off with time, and, of course, there are significant side effects.

Study of rheumatoid arthritis has taught us a lot about how we approach MS. Like MS before the modern treatment era, rheumatoid arthritis was an incurable, disabling, life-shortening disease. The major breakthrough occurred in the 1990s with the introduction of anti-TNF therapy by Marc Feldmann and Ravinder Maini. Tumor necrosis factor, or TNF, is a mediator of inflammation in the joints. Anti-TNF drugs (with billions of dollars in sales) are monoclonal antibodies that neutralize TNF, such as Remicade, Enbrel, and Humira. Other monoclonal antibodies have been found effective in rheumatoid arthritis, especially ones that target a second inflammatory mediator in the joint, IL-6. Interestingly, TNF is also elevated in the brains of patients with MS, but it turns out that anti-TNF treatment makes MS worse. In fact, MS develops in some rheumatoid arthritis patients who are treated with anti-TNF drugs. Nonetheless, despite all the new agents for rheumatoid arthritis, progress has stalled over the last decade, and although rheumatoid arthritis patients are doing far better, total disease-free status is not easy to obtain. The answer may lie in combination therapy and, of course, treating as early as possible.

The challenge we face with MS is that, in many instances, the disease has been present for a number of years before the diagnosis is made. We can usually determine this by MRI imaging. In fact, there are studies suggesting that if treatment is delayed, patients do not do as well over time. In a study of the subcutaneous beta interferon drug Rebif for relapsing MS, there were two stages of treatment. In the first stage patients were placed on Rebif or placebo for a two-year period. The results were clear: those on Rebif had fewer relapses, less disability, and less activity on MRI. In the second stage, those who were initially on the placebo were then given the drug and were followed for another two years. This was called the delayed treatment group. It was found that at four years those who had been continuously treated with Rebif fared better than the delayed treatment group. Moreover, a further follow-up at eight years confirmed a better outcome in those continuously treated as opposed to those who received delayed treatment. This would argue that ground lost when treatment is delayed cannot be easily made up and if we are to cure MS, we need to treat at the first sign of disease.

In a landmark study conducted at the Queen Square Hospital in London, patients who presented with their first MS neurological symptom between 1984 and 1987 underwent MRI imaging and were then followed and analyzed at five-year intervals over a twenty-year period. The first neurological symptom of MS that we can recognize clinically is called clinically isolated syndrome (CIS). Examples of CIS include loss of vision, numbness and tingling in an extremity, and incoordination. Some patients with CIS already have signs of MS on their MRI, others minimal if any. Two crucial questions related to CIS are: What happens to patients with CIS if they are not treated? and, What is the effect of treatment on those with CIS? As it turns out, virtually all of the patients in the Queen Square study remained untreated during the course of follow-up, something that could not happen today, as it is now known that early treatment has a positive effect and virtually all patients are now treated. In the Queen Square study, MS developed in 88 percent of those with abnormalities on their MRI at the time of CIS, whereas it developed in only 19 percent of those with a normal MRI. The volume of MRI abnormality and changes during the early time points correlated with the amount of disability accumulated twenty years later. Furthermore, the rate at which MS progressed on MRI was three times higher in those who developed secondary progressive MS than in those who remained relapsing-remitting. These findings suggest that the MRI might identify patients who should receive more aggressive treatment at disease onset.

A similar result was obtained in a Barcelona study of 175 CIS patients followed for thirteen years, which demonstrated that patients with increased numbers of MS spots on MRI while on interferon beta therapy accumulated more disability. As clinicians, we are often confronted with a dissociation between the number of spots on MRI and the degree of disability, perhaps related to an individual's age and the degree to which the brain can compensate. However, there is no disconnect between the degree of MRI involvement and prognosis over time.

If we can identify MS at the time of CIS, should patients with CIS be treated? Because some patients do well without treatment (although this may create a false sense of security as patients are not followed long enough), some physicians argue that it is better to be off therapy or not on strong therapy until the disease declares itself. However, most physicians feel waiting is too late. One can only cure MS by stopping it in its tracks. Trials of interferons and Copaxone show that treatment at the time of the first clinical manifestation of MS delays developing clinically definite MS. Furthermore, in a study in which CIS patients were treated for two years with placebo or interferon (Betaseron), those on treatment did better. In addition, those who didn't have a two-year delay in treatment were better off clinically eleven years later.

With the advent of MRI imaging, we sometimes identify MS before it presents clinically. This happens when a person has an MRI for another reason (headaches, head trauma) and MS is discovered. This is called radiologically isolated syndrome (RIS) and is the very earliest we can see MS. After the initial description of MS by Charcot in 1872 but before there were treatments for MS, some patients with MS who did not develop disability were only discovered to have had MS as an incidental finding at autopsy.

In diseases such as breast cancer, there are genetic factors and biologic features of the tumor that have prognostic implications and determine the type of treatment that will be given. We don't have such strong genetic factors linked to MS prognosis, but there are poor prognostic features in early MS, including number of relapses in the first two to five years, multiple lesions on MRI, attacks affecting the motor system, and being African American. In our CLIMB study we found a lower rate of brain atrophy over time in patients with a benign disease course in which no disability developed. Understanding benign forms of MS could give important clues to the underlying process in the disease. It's possible that people with benign MS have a specific protective factor that could be identified, or they may be missing an important pathogenic factor. In HIV AIDS there is a subcategory of patients who don't become ill, called long-term non-progressors. A number of genetic and other protective factors have been identified as being linked to long-term non-progressors for HIV; they are not that well defined for MS.

Important insights into MS can be learned from the occurrence of MS in children because it provides a window into the earliest events that trigger MS.

Although MS generally begins in young adulthood, pediatric MS is a well-recognized entity. In 1982, Steve Hauser and I studied immune abnormalities in children with MS ages three, five, and ten. Pediatric MS is divided into prepubescent children (age two to ten) and adolescents whose disease onset is before age eighteen. The female predominance of MS is not seen in prepubescent children, consistent with the fact that female hormones contribute to MS susceptibility. MS is three times more common in females than in males. Pediatric MS is primarily a relapsing inflammatory disease; there are few cases of progressive MS in children. Most of the drugs used for adult MS are effective in pediatric cases.

MS DRUG DISCOVERY

As discussed previously, the first drug approved for MS, in 1993, was a beta interferon called Betaseron. It was actually tested for the wrong reason based on the theory that MS is caused by a virus. Interferons are antiviral agents. We now know that interferons help MS through their immunologic effects, but their action has nothing to do with combatting viruses. The second drug to be approved for MS, in 1996, was also a beta interferon, called Avonex. Larry Jacobs, the neurologist who pioneered the development of Avonex, was interested in its antiviral effects. Indeed, he first planned to test the drug in ALS on the theory that ALS was caused by a virus. When he couldn't find enough ALS patients to test, he turned to MS and injected the drug into the spinal canal. Avonex was ultimately developed and marketed by Biogen. Interestingly, before they developed the beta interferon Avonex, they tested a gamma interferon, also because of its antiviral properties. The only problem was that unlike beta interferon, gamma interferon is a type of interferon that has a strong stimulatory effect on the immune system. I remember being at a Biogen advisory board meeting where gamma interferon was being discussed. The immunologists in the room warned against giving gamma interferon, arguing that if MS was an immune-mediated disease, gamma interferon would make MS worse, but the virologists liked the idea of gamma interferon because it was a recombinant protein made by genetic engineering. Biogen decided to test it in MS. It was one of the few drugs tested in MS that induced MS attacks, further proof that the immune system was driving the disease and it was important to dampen the immune system in MS, not stimulate it.

The third drug approved for MS, in 1996, was Copaxone (glatiramer acetate). It was developed at the Weizmann Institute in Israel and was based on the animal model of MS, experimental allergic encephalomyelitis (EAE). Proteins are made of building blocks called amino acids; Copaxone (initially called copolymer-1) was a random mixture of four amino acids designed to mimic a brain protein called myelin basic protein, or MBP.

It is possible to induce EAE in animals by injecting them with MBP mixed in an adjuvant that increases the immune system response. Injecting animals with MBP in an adjuvant induced autoimmune disease–causing T cells specific for MBP. They migrated to the brain, where they recognized MBP on the myelin sheath and caused inflammation and paralysis. The scientists wanted to see if injecting copolymer-1 with an adjuvant would mimic MBP and cause disease. The experiment failed. The T cells induced by injecting copolymer-1 with an adjuvant did nothing. However, the scientists turned the experiment around and asked whether they could use the copolymer as a treatment and whether injecting the copolymer alone could ameliorate the disease. This time the experiment worked, and Copaxone went on to become one of the best-selling drugs for the treatment of MS. The mechanism by which Copaxone was postulated to work, however, turned out to be wrong. Copaxone didn't act by mimicking myelin basic protein; it worked by conditioning immune cells called monocytes. It was also later tested for possible positive effects in animal models of Alzheimer's disease and ALS. Furthermore, as we will learn, the basic principles related to EAE would play a critical role in discontinuing a vaccine for the treatment of Alzheimer's disease.

After the FDA approval of Avonex and Copaxone in 1996, the era of MS therapeutics was launched, and doctors had three drugs for the treatment of MS, all given by self-injection. At this time, patients and doctors referred to MS drugs as the ABC drugs (Avonex, Betaseron, and Copaxone). Rebif, another form of beta interferon, given at a higher dose than Avonex, was approved in Europe in 1998. It was then tested in a head-to-head comparison against Avonex and was shown to have better efficacy in reducing MS attacks. Based on this, Rebif was approved in the United States in 2002, and we then had the ABCR drugs for the treatment of MS.

The trial of Rebif versus Avonex was the first of many direct comparisons of MS drugs. With four drugs on the market at a cost to the user of approximately $10,000 per year, it became clear that having an approved MS drug on the market was extremely lucrative—they do a billion dollars per year in sales. MS physicians such as myself were courted by the pharmaceutical industry to be on speaker bureaus, act as consultants, and advise how a company could best market and promote a particular drug. Close ties developed between some physicians and a particular drug company, and these physicians were known to prescribe one drug over another. Because of reporting laws, each company could know which drugs were prescribed by a particular physician. On more than one occasion, drug company sales reps asked me why I didn't prescribe more of their drug. Advisory board meetings in nice locations and dinners in fancy restaurants became commonplace. Over time, however, stricter rules were put in place to limit attempts by pharmaceutical companies to influence the prescribing habits of physicians. At a recent meeting, I was asked to bring my own lunch, as rules prohibited the company from providing me with a meal. I brought a tuna fish sandwich.

A major event in drug treatments of MS occurred in 2004 when the FDA approved the Biogen drug Tysabri for relapsing forms of MS. Of all the drugs developed for the treatment of MS, Tysabri followed the most logical path of development. The drug was developed directly from the EAE animal model of MS, and its mechanism of action established the importance of cells moving from the bloodstream into the brain to combat MS. It was one of the most effective drugs approved for the treatment of

MS. Unfortunately, it also had a completely unexpected lethal side effect, which has limited its use in MS and has plagued other MS drugs as well.

The development of Tysabri began with experiments in which scientists investigated how white blood cells (T cells) move out of the bloodstream and into tissues. They found that in order to leave the bloodstream, the T cell must first bind to the vessel wall. There are specific homing receptors on the surface of T cells that bind to the blood vessel wall; if the receptors are blocked, the T cell is trapped and can't leave the bloodstream and will not eventually cross the blood-brain barrier.

The receptor used by the T cells is called alpha 4 integrin. Remember, EAE is caused by T cells that leave the bloodstream, enter the brain, and attack the myelin sheath. The identical process is postulated to happen in MS. T cells entering the brain cause an MS attack. Thus, if T cells are trapped in the bloodstream and can't enter the brain, there should be no MS attacks. When Tysabri was given by intravenous infusion once per month, it reduced MS attacks by 68 percent, far better than the 35 percent that was observed with the ABCR injectable drugs. It also had a greater effect on disability development and MRI changes than the ABCR drugs. Within two months of the approval of Tysabri, we had treated two hundred patients in our MS Center with the drug. We were ready to begin treating almost all our relapsing patients with Tysabri when disaster struck. Two months after the approval of Tysabri, two patients who were being treated with Tysabri developed a fatal brain infection called progressive multifocal leukoencephalopathy, or PML, and Tysabri was withdrawn from the market.

PML is an extremely rare brain infection that occurs in people with a suppressed immune system, including those with AIDS or cancer and people receiving immunosuppression for organ transplantation. PML is caused by a virus called the JC virus. The JC virus was discovered in 1971 in the brain of John Cunningham, a patient with Hodgkin's lymphoma, and named for him. More than 50 percent of the adult population is infected with the JC virus. In healthy adults, the infection is lifelong but does not cause disease. The rarity of PML compared with the ubiquitous nature of

JC virus infection indicates that there are barriers to developing PML, the most important being the immune system. PML results from reactivation of virus in the body, not a newly acquired infection. The disease is diagnosed by MRI and a spinal tap.

Because of Tysabri's effectiveness, when it was voluntarily withdrawn from the market by Biogen, MS advocacy groups successfully lobbied the FDA to have it put back on the US market in 2006 after reviewing two years of safety and efficacy data. No other brain infections had occurred.

Patients taking natalizumab were required to enter a registry program called TOUCH for monitoring. When natalizumab was reapproved in 2006, the risk of PML was estimated at 1 in 1000 over two years of treatment (it is rare for PML to occur after only one year of treatment). Over the next several years, three risk factors for PML were identified: treatment for more than two years; prior immunosuppressive therapy; and the presence of antibodies against the JC virus, the pathogen that causes PML.

The development of the JC virus test was a major advance in the use of Tysabri in patients with MS. People with low antibody titers and no previous immunosuppressants have only a 1 in 8,000–10,000 risk of developing PML, and doctors are thus comfortable in treating these patients with Tysabri. If a person is positive for the JC virus and has been on immunosuppressant drugs, it is 1 in 100. Some patients elect to continue on Tysabri despite the risk and being JC virus positive. I have a patient who thinks it is worth it. Since the introduction of Tysabri, over 200,000 MS patients have been treated with the drug with an overall incidence of PML of approximately 4 per 1000. By the end of 2020, there were over eight hundred cases of PML in people taking Tysabri; one-third have died.

It is now standard practice to obtain an MRI and perform antibody testing every six months in patients taking Tysabri. Although PML is especially associated with Tysabri, there have been cases reported with newly approved MS drugs such as Tecfidera and Gilenya, although incidence is much rarer. PML has been observed in patients with the autoimmune disease lupus who have been treated with the drug rituximab. This is relevant because a drug like rituximab has also been approved for MS. A drug called Raptiva was marketed to treat psoriasis and then withdrawn from the market because of PML. Like Tysabri, it also affected white blood cells traveling through the body. It is probable that if there were no PML with Tysabri, almost all patients with relapsing MS would have been treated with the drug.

Following the widespread use of the ABCR injectable drugs, the race was on to develop a drug that could be taken orally so people wouldn't have to inject themselves. Over the next decade five oral drugs would be developed. Four would make it to market. The story of these oral drugs provides a window into the world of how drugs are developed by pharmaceutical companies and sheds further light on the disease mechanisms that drive MS. Any time a drug is shown to help a disease, we learn more about the disease process.

ORAL DRUGS FOR MS

The first oral drug to be approved for the treatment of MS was Gilenya, or fingolimod, developed by Novartis. Prior to introducing Gilenya, Novartis had developed another interferon for the treatment of MS called Extavia, which was approved in 2009. Gilenya was approved a year later in 2010. The development of Gilenya for the treatment of MS began in Japan with drug discovery in the soil. We know that nature was the first source of medicines to treat human disease. Aspirin was derived from a substance in willow bark, and Alexander Fleming discovered a fungus that produced a bacteria-killing substance, which turned out to be penicillin.

Novartis was created from a merger of Ciba-Geigy and Sandoz in 1996. One of the more successful Sandoz drugs was an immunosuppressant called cyclosporine, used in kidney transplantation to reduce the risk of organ rejection. Like Gilenya, the development of cyclosporine followed the discovery of new strains of fungi isolated from the soil in 1970. Cyclosporine was approved for use in human transplantation in 1983. Because it inhibited the activity of T cells, which were thought to be at the center of MS, in 1990 Sandoz tested cyclosporine at twelve MS centers across the country in patients with primary progressive MS. Although there were some positive effects, there were many dropouts because of renal toxicity. Cyclosporine was abandoned as a treatment for MS.

Ironically, two decades later, Novartis would have success in MS with another drug developed from the soil, fingolimod. Tetsuro Fujita, a Kyoto University pharmacology professor, was investigating plants used in traditional Asian medicine. For centuries Chinese medicine practitioners maintained there were health properties in an Asian fungus that invades and destroys insects. Fujita postulated that an immunosuppressant chemical would be present in the fungus, which attacks insects in the winter with its chemical arsenal. By summertime, the insect is dead, and its corpse becomes a place for the fungus to grow. Fujita and his team analyzed various fungi and found a substance that acted like cyclosporine. They modified the chemical structure to make it less toxic and tested it for its ability to delay skin graft rejection in rats. From these experiments, fingolimod was born, and in 1997 Novartis acquired rights to the drug. They tested fingolimod in Phase 2 and Phase 3 clinical trials for renal transplantation but found it no better than available drugs, including cyclosporine. Rather than discard the drug, scientists in the lab tested fingolimod in the EAE animal model of MS and found striking results. Based on this, a development plan for fingolimod in both relapsing and progressive MS was initiated.

In animal models and in humans, fingolimod was found to reduce white blood counts, affecting immune cells in the blood. It was postulated that fingolimod might accelerate the movement of lymphocytes out of the bloodstream and into lymph nodes, although the mechanism was not known. Ultimately, it was discovered that the target of fingolimod was a receptor called the S1P receptor. Fingolimod caused S1P receptors to move from the surface of the cell to the inside of the cell, which in effect blocked S1P receptor function. This caused autoimmune disease–inducing T cells to accumulate in the lymph node.

Although there are other actions of Gilenya (fingolimod), it is believed that the main mechanism of its action is trapping cells in the lymphoid tissue. Thus, Gilenya is similar conceptually to Tysabri, in that it traps cells outside the brain so they can't enter the brain and cause damage. Tysabri blocks the ability of cells to leave the blood vessel and enter the brain; Gilenya doesn't allow the lymphocytes to enter the bloodstream. It is difficult to study the immune system in patients on Gilenya because there are decreased white blood cells in the blood.

Like Tysabri, the proof that Gilenya had an effect in relapsing MS was a six-month, placebo-controlled Phase 2 trial involving 281 patients. Patients on treatment had a lower number of gadolinium-enhancing lesions on MRI and fewer relapses compared to placebo. Following this, Gilenya was evaluated in a two-year double-blind Phase 3 study known as FREEDOMS, which involved 1,272 patients with relapsing-remitting MS. Two doses were tested, 0.5mg and 1.25mg. The results were clearly positive, and patients on treatment had fewer relapses and less disability. On MRI imaging, treated patients had fewer new or gadolinium-enhancing lesions and less brain volume loss or atrophy.

As new medications were tested in MS, the question of how they compared in efficacy to approved medications was raised. A head-to-head trial of the higher-dose interferon Rebif versus Avonex led to approval of Rebif in the United States. Novartis also performed a head-to-head trial comparing Gilenya to Avonex. Because one drug was given orally and the other by intramuscular injection, it was designed as a double-blind, double-dummy Phase 3 study. Thus, all patients took a pill or placebo pill every day and an interferon injection or placebo injection once a week. The trial was called TRANSFORMS and involved 1,292 patients with relapsing-remitting MS. The results showed that the relapse rate, along with measures of MS, was lower in those receiving Gilenya versus Avonex as shown by MRI imaging (number of new or enlarged lesions, number of gadolinium-enhancing lesions, and brain-volume loss at twelve months).

Because of side effects associated with Gilenya, patients have three conditions monitored when therapy is initiated. First, the presence of S1P receptors in the heart causes a transient and generally asymptomatic reduction of heart rate after the first dose. Because of this, the first dose of Gilenya is given under medical supervision over a six-hour period. Another side effect of Gilenya is swelling of the macula in the eye, so patients have a baseline eye exam before starting treatment. Finally, because of increased risk of shingles, patients who haven't had chicken pox must receive a vaccination before treatment. Although there is a slightly increased incidence of respiratory infection with Gilenya, there don't appear to be significantly more infections. Rare cases of PML have been reported in patients taking Gilenya. To date, nine cases have been reported in 160,000 patients dosed. The combined data from the Phase 3 trials did not suggest an increased incidence of malignancies associated with fingolimod treatment. This is an important point, because as we will see in our discussion of brain tumors, one of the ways in which cancer promulgates is to suppress the immune system so the immune system can't attack the cancer. Thus, there is always the theoretical risk of more cancer in people who are taking immunosuppressant drugs. With the success of Gilenya, companies including Novartis, Celgene, and Actelion are developing second-generation S1P inhibitors that are designed to have fewer cardiac side effects and a shorter half-life. In 2018, ten years after the approval of Gilenya for relapsing MS, a multicenter randomized study headed by Tanuja Chitnis from our center successfully compared Gilenya to the interferon Avonex, and Gilenya became the first FDA-approved drug for treatment of pediatric MS.

The second oral drug to be approved for MS was Aubagio (teriflunomide) marketed by the company Genzyme. Like so many stories in the development of drugs for MS and other diseases, Aubagio's path was tortuous and took years, involving different companies and corporate mergers.

In the 1990s, a company called Aventis developed a drug called leflunomide as a treatment for rheumatoid arthritis, under the brand name Arava. Because rheumatoid arthritis falls into the category of autoimmune diseases, the theory was that a drug that worked for rheumatoid arthritis would benefit patients with MS. Researchers at Aventis modified Arava to remove some of its potential side effects and created a new drug, ultimately called Aubagio, which had the potential to become the breakthrough oral drug for the treatment of MS that people were searching for. While Aventis was developing Aubagio, Sanofi bought Aventis; then, seven years later, Sanofi bought Genzyme, which had the powerful monoclonal antibody MS drug Lemtrada. After the purchase, Sanofi gave Genzyme the MS portfolio. If Sanofi had not acquired Genzyme, then Genzyme would not have had Aubagio.

Like all FDA-approved drugs for relapsing-remitting MS, Aubagio works on the immune system. Aubagio is effective because it quiets an overactivated immune system. When immune cells are stimulated, they divide and spring into action. In an autoimmune disease such as MS, the response is excessive, and activated lymphocytes proliferate. Aubagio works by stopping lymphocytes' ability to divide.

Three separate Phase 3, placebo-controlled trials showed the effectiveness of Aubagio in relapses, MRI measures, and disability, leading to FDA approval. The TEMSO trial enrolled 1,088 patients, who were randomized to receive two doses of Aubagio versus a placebo. A second study called TOWER involved 1,069 patients randomized in a similar way. Finally, the TOPIC study investigated 618 patients with a clinically isolated syndrome—the earliest MS is recognized clinically—and their risk to convert to clinically definite MS. Aubagio was easy to take. The major side effect related to risk of genetic effects during pregnancy, which made it less likely for Aubagio to be used in women of childbearing age. There was also a slight risk of hair thinning. Later, a trial compared Aubagio to injectable beta interferon and found no difference between the treatments, apart from their delivery, as one had to be injected and the other was oral.

The third oral drug approved for relapsing MS was Tecfidera (BG-12), developed by Biogen. The FDA approved Tecfidera in 2013. There are many pathways for the development of drugs. Some come from animal models of a disease, others from the discovery of a specific gene or pathway that is then applied to the disease. In the case of Tecfidera, the journey began a decade earlier with treatment of a small number of MS patients in Germany with a drug used for treating psoriasis.

The German drug was called Fumaderm. Serendipitously, it came to the attention of Biogen when Akshay Vaishnaw, the lead medical director in the company's dermatology program, traveled to Germany to evaluate a drug for psoriasis. Biogen had a drug called Amevive that had been approved for dermatology, and the company was looking for other drugs for their dermatology franchise. Vaishnaw heard from a dermatologist in Germany that the most common oral medication for psoriasis was Fumaderm, which had been approved only in Germany. As he was evaluating Fumaderm for psoriasis, he was told that a neurologist who had MS took Fumaderm and felt he had a positive response. The neurologist proceeded to do a pilot trial on ten patients with relapsing-remitting MS. It was an open label MRI study in which all patients knew they were taking the drug. He took patients who had active disease, as evidenced by gadolinium-enhancing lesions on MRI, treated them with Fumaderm, and found a dramatic reduction in the lesions. The most common side effects of Fumaderm were gastrointestinal symptoms and flushing.

Vaishnaw brought the data to Al Sandrock, head of drug discovery at Biogen. Sandrock was skeptical, especially because the trial involved only ten patients and there was no control group. However, as it turned out, Sandrock had been working with Henry McFarland at the NIH, who had been conducting monthly MRI scans on his MS patients and was developing the concept that one could do a single arm study without a control group in relapsing-remitting MS if one compared pre- versus post-treatment scans. Sandrock carried out a series of statistical simulations and concluded that a 25 percent decrease in the average in the number of enhancing lesions could occur by chance. Above 25 percent was unlikely to be due to chance, and above 50 percent was very unlikely to be due to chance. The Fumaderm data showed a greater than 50 percent reduction in lesions. Sandrock also reasoned that psoriasis, like MS, was a type of autoimmune disease. He was ready to bring the product to Biogen, but first, Sandrock asked Gilmore O'Neil and Mike Panzara, colleagues of his at Biogen, to look at the data and confirm that the MRI results showed such a dramatic effect. When they reported that the data were real, Sandrock went directly to a 257-patient placebo-controlled dose-finding Phase 2 study in which patients were treated for two years. It was a bold move, but Biogen was in need of an oral MS drug. The decision paid off. The company found a dose-dependent reduction of MRI lesions and a 32 percent reduction in MS relapses. As it turned out, Biogen ended up acquiring Fumapharm, the company that made fumaderm.

Excited by the Phase 2 results, they performed two Phase 3 studies. Phase 3 trials are large, expensive undertakings. Because the FDA usually requires two clinical trials to approve a new drug, two independent trials were carried out. In the first, two doses of what would ultimately be named Tecfidera were compared against placebo. In the second trial, Tecfidera was compared against both placebo and Copaxone. The trials were performed at two hundred sites in twenty-eight countries. Imagine the logistics. Biogen hadn't tested twice-daily Tecfidera in their Phase 2 trial but felt it was worthwhile to establish a dose effect. To their surprise, twice a day was just as good as three times a day. Furthermore, twice a day had better efficacy than interferon. The reduction in the relapse rate relative to placebo approached 50 percent. Although the exact mechanism of action of Tecfidera in MS is unknown, it clearly works by affecting the immune system. Tecfidera induces immunomodulation on different immune cell types and may act through Nrf2, a regulator of cellular resistance to oxidants.

The development of Tecfidera began with dermatology and ended up in MS. A great deal of serendipity was involved. Biogen hadn't done dermatology before. If Akshay hadn't brought it to Al Sandrock's attention, they would never have considered the drug. Tecfidera was approved by the FDA in 2013, a little over a decade from the time Akshay first was made aware of its effects in a group of ten MS patients in Germany.

The successful development of Tecfidera for MS did not come from the laboratory, as did Copaxone and Tysabri. It came from empirically testing a drug with anti-inflammatory effects in people with MS and finding a positive effect. Tecfidera has been shown to be effective in the EAE animal model of MS, but this testing was done *after* it was tested in people.

Two other therapies participated in the race for the orals.

The first was laquinimod, developed by Teva Pharmaceuticals, the maker of Copaxone. Laquinimod showed positive effects in Phase 2 trials in relapsing MS but unfortunately did not succeed in Phase 3 trials. The second was cladribine. In 2010, Phase 3 trials of both cladribine and Gilenya were published in the same issue of the *New England Journal of Medicine*. Gilenya went on to become the first oral drug approved for MS, whereas cladribine was met with rejection from the FDA and EMA (the European regulatory agency) after the agencies detected an increased risk for cancer and some infections. However, continued analysis of the results, including a follow-up article published in 2015, allayed concern about these side effects and showed no evidence of a higher cancer risk in patients taking cladribine.

In 2017, cladribine, now called Mavenclad, became the fourth oral medication approved for relapsing MS first in Canada and Europe, and then in the United States. Mavenclad differs from the other oral medications as it is taken in two weekly courses a year apart, whereas the other oral medications are taken once or twice every day. Aside from the convenience, taking Mavenclad so infrequently allows these patients to avoid a daily reminder of their illness.

Like all drugs for relapsing MS, Mavenclad acts on the immune system and works by decreasing immune cell activation. One finding in the Phase 3 trial of Mavenclad was the large number of patients who had NEDA or no evidence of disease activity as measured by relapses, disability, or MRI changes. With more effective drugs for relapsing MS, NEDA is becoming a new measure of response to therapy in MS.

THE B CELL SURPRISE

The development of treatments for MS began in the '90s with injectable drugs that modulated the immune system, followed by Tysabri infusions, which blocked movement of cells into the brain. In the past decade, oral medications were developed that also modulated the immune system.

More recently, a major shift has occurred in our understanding of the immune system in MS with the demonstration that targeting immune cells called B cells has a dramatic effect on decreasing MS attacks. There are two major branches of the immune system—humoral and cellular. The humoral branch involves antibodies, made by B cells, and the cellular branch, which involves white blood cells called T cells. As discussed, MS and the EAE animal model of the disease have always been considered a T cell process. The discovery of the B cell as a target in the fight against MS has resulted in a paradigm shift in thinking about the disease and its treatments.

Steve Hauser was one of a number of scientists who focused on B cells early on. In addition to Steve, Anne Cross at Washington University in St. Louis and Amit Bar-Or in Montreal conducted the initial studies on B cells in MS. Anne was recently awarded the MS Dystel Prize for her work on B cells, but Steve has been most responsible for bringing therapy directed at B cells to the clinic, led by his long-standing interest in the role of antibodies in MS. By 1995, after a decade of work, Steve and his team had developed the ability to produce an MS-like condition in a New World marmoset, a primate about the size of a guinea pig. The condition mimicked MS pathologically and involved a humoral immune response. He found antibodies made by B cells were required to damage myelin in marmosets. Moreover, B cells can be detected in MS lesions.

Beginning in 1999, Steve and his team turned to a B cell–based therapy using Rituxan. Rituxan was a drug approved for treating non-Hodgkin's lymphoma, chronic lymphocytic leukemia, and rheumatoid arthritis. It worked by depleting B cells that have the CD20 molecule on their surface. Steve wanted to see if depleting these B cells had any effect on MS. He was especially interested in whether Rituxan affected excess antibody in the cerebrospinal fluid (CSF), called oligoclonal bands, in MS patients. When MS spinal fluid is placed on an agarose gel and stained for protein, the excess antibody forms discrete bands; the presence of these bands has been used to diagnose MS. In the past, all patients underwent a spinal tap to look for oligoclonal bands indicating inflammation in the central nervous system. A hallmark of MS is the presence of these bands in the CSF but not the blood. These bands are observed in 95 percent of MS patients.

Steve initially had trouble getting support for the trial. The NIH turned down a grant proposal from his team in 2001. According to Steve, the comments on the grant application reflected profound skepticism that humoral B cell immune mechanisms could be central to MS. The skepticism about a B cell–focused therapy was widespread at the time. The general consensus in academia was that MS was caused by T cells, so pursuing B cells was a waste of time. People did not believe the increased antibody in the spinal fluid, as indicated by oligoclonal antibodies, was important.

After being turned down by the NIH, Steve pitched Genentech, which held the patent on Rituxan. He was eventually able to convince the pharmaceutical company to perform a yearlong MS trial with Rituxan, although their experience showed that antibody levels were unaffected by treatment with the drug. Concerned that some patients would be on a placebo for a year, the FDA shortened the study to six months, so only a single course of Rituxan was given. Even so, the results were dramatic.

In the prestigious Charcot Lecture, Steve described how he felt in 2006 when the results of the study were unblinded: "The joy was twofold. First, we saw evidence of a potentially powerful new approach for therapy for relapsing MS. Second, despite this success, it was also evident that our scientific rationale behind the clinical testing of Rituxan for MS was almost certainly incorrect. Looking back, this was the best possible result of a translational research experiment. We identified a novel approach that appeared to offer significant benefits for patients, and yet the data also sent us back to the laboratory in new and unexpected research directions."

In this short study, the investigators found an almost 91 percent reduction in gadolinium-enhancing spots on MRI. The positive results observed by Steve and his group led to additional trials with Rituxan and later with Ocrevus, a monoclonal antibody that targeted the same structure on B cells as Rituxan.

Following publication of the initial study in 2008, which showed Rituxan's dramatic effect in relapsing MS, Genentech's patent was running out on the drug. They switched to the drug now marketed as Ocrevus for the Phase 2 trial. Like Rituxan, Ocrevus targets B cells with CD20 molecules on the surface. Genentech argued the switch was beneficial because Ocrevus was a totally humanized molecule. By contrast, Rituxan was an antibody composed of human and mouse parts, raising the risk the patient's immune system would develop antibodies against the medicine. The Phase 2 trial of Ocrevus, published in 2011, showed positive effects that were as good as Rituxan, as did the Phase 3 trial, published in 2016 in the *New England Journal of Medicine.*

There were two remarkable findings with Ocrevus. First, it had a dramatic effect in relapsing MS. For example, the drug is 99 percent effective in stopping new or enlarging spots on MRI, meaning it stops 99 percent of white matter disease. Second, it worked in primary progressive disease, thus becoming the first FDA-approved drug for progressive forms of MS. Indeed, the positive results in progressive disease led to fast-track approval in 2017 and breakthrough therapy designation by the FDA.

Unfortunately, the results in primary progressive MS were misleading. Although a previous study of Rituxan in primary progressive MS did not succeed, subgroup analysis suggested younger patients or patients with inflammatory components may have responded. Thus, patients were carefully selected for the Ocrevus primary progressive trial who were younger patients, below the age of fifty-five, with a disability as measured on the

Expanded Disability Status Scale (EDSS) of less than 6.5, and a duration of disease less than fifteen years (an EDSS of less than 6.5 means that a patient can walk without needing bilateral support such as a walker). That meant the primary progressive trial did not include patients with longer-standing progressive MS, nor did it include older patients.

When we talked about the drug, Steve said he was worried about misrepresenting or hyping the progressive results, because they were incremental at best. Given FDA approval of the drug and the ability of pharmaceutical companies to market directly to consumers, this could create a problem. Nonetheless, we know that there are subgroups of patients and that treatment as early as possible is advised. This suggests treatment at the earliest stages of progressive disease may be crucial.

Steve told me he is excited about the possibility of treating patients at "the first whiff of disease," with an easy-to-administer and very effective drug such as Ocrevus. He believes the drug has the potential to turn off MS for a very long time. Of course, it will take years to learn whether shutting down early inflammation really does prevent later progression, because underlying progression of disease or silent progression can occur although inflammation has been shut down. I always wanted to treat MS with cyclophosphamide at first sign of MS but couldn't do it because of the toxicities associated with this chemotherapy agent.

Although Ocrevus is very promising, it could be derailed by unexpected side effects. The drug does not appear to carry an increased risk for progressive multifocal leukoencephalopathy (PML), the rare brain disease that temporarily stopped Tysabri in its tracks. It was initially thought that there was a slightly increased risk of breast cancer, though this has not been borne out by further studies. One reason Ocrevus appears to be safe is that it kills only a subset of B cells. Although the CD20 structure is expressed by all B cells and Ocrevus binds to CD20 positive B cells everywhere in the body, it kills only the B cells that are moving outside the body's lymph nodes. Only 2 percent of B cells are in the blood, so the vast majority of B cells are not killed by the treatment. Nonetheless, B cells are needed to make antibodies, and we don't know whether COVID-19 vaccines will work as well in MS patients taking Ocrevus.

Steve Hauser and his team were interested in Rituxan—and later Ocrevus—because they were focused on antibodies. Unexpectedly, the drug appeared to act on B cells themselves.

"In some ways, it was luck that the trial worked," Steve told me years later.

The stunning success of the Rituxan trial sent Steve and others back to the laboratory to investigate why it worked so well. One of the beautiful things about science is that there is no fake news. All biologic observations are, by definition, real and must have an explanation. B cell–targeted therapy worked against MS. But if MS is a T cell disease, why did Ocrevus work?

We now know treating people with therapy against B cells causes a number of changes, including in regulatory B cells. There are pro-inflammatory B cells and anti-inflammatory B cells, and the pro-inflammatory B cells are decreased. The discovery of the role of B cells now opens the path for measuring B cells as biomarkers and to further develop drugs to target B cells. One such drug, Kesimpta, developed by Novartis, was also studied by Steve Hauser and Amit Bar-Or with positive results of two large Phase 3 trials published in the *New England Journal of Medicine* in 2020. It performed better in comparison to Aubagio, an oral approved drug for relapsing MS. Kesimpta is given subcutaneously once a month and received FDA approval for the treatment of relapsing MS. However, another drug targeting B cells, named Ataticept, actually made MS worse. It appears Ataticept works in the wrong set of B cells. Science is never as simple as it seems.

The approval of Ocrevus was truly a landmark event, and large numbers of patients are now being treated. Interestingly, the health systems in Scandinavian countries are paying for Rituxan rather than Ocrevus, because Rituxan is less expensive.

The game-changing nature of B cell therapy raises a central question about the nature of MS. Does the success of Ocrevus mean MS is, in fact, a B cell disease? Nearly half a century after researchers concluded MS was a T cell disease, the question would be debated on one of the MS world's largest stages, the 2017 European Council for Treatment of MS (ECTRIMS), at the Palais des Congrès in Paris. Once a small group gathering, the ECTRIMS conference has become an industry-driven affair with ten thousand doctors in attendance. Slick marketing adorns the giant halls touting various MS drugs, and large overhead screens project company videos about their drugs. Researchers also present the latest scientific advances in MS clinical trials.

We cannot develop drugs without interacting with the pharmaceutical industry, and I must admit it is extraordinarily exciting to see ten thousand doctors in one hall learning about the latest discoveries in MS. There is also marketing in science, in an attempt to apply popular culture to science to make it more accessible. This was on full display at the ECTRIMS in Paris as I prepared to listen to a lecture dubbed "Rumble in the Jungle," named for the most famous boxing match of all time, the 1974 Ali-Foreman fight in Kinshasa. Steve Hauser and David Hafler were debating whether MS is a B cell disease or a T cell disease. Steve took the position that MS is a B cell disease. David argued MS is a T cell disease. Each debater had fifteen minutes to make his case. I took pride in seeing two of my first fellows on a giant stage, now professors and heads of neurology departments, and major figures in the study of MS.

Steve began his talk with a picture of me undergoing plasma exchange in 1981 with him and David Hafler standing at my side as I sat in a special chair, IV needles in each arm. By undergoing plasma exchange, I was following in the long tradition of physicians who have treated themselves, from cardiac catheterization to taking *Helicobacter* bacteria to cause ulcers.

David said T cells initiate MS when T cells that normally secrete the anti-inflammatory chemical IL-10 begin to secrete pro-inflammatory chemicals IL-17 and interferon gamma. Steve countered that B cells outnumber T cells in MS lesions. Despite the pugilistic billing, the two sides of the "Rumble in the Jungle" debate were more in agreement than not. Steve, who was taking the position that MS is a B cell disease, concluded, "B and T cells are inseparable allies. They cannot exist apart, and in MS, they are partners in crime, but B cells are the ringleader." For his part, David concluded, "MS is not a T cell disease. MS is a total immune system disease that involves all aspects of the immune system."

There are dozens of elegant studies showing how T cells change in the disease. Independent of this, however, we now know that B cells indeed sit at the heart of MS. In fact, if you look at older, approved therapies for MS, they all affect B cells in some way. Interferon and Copaxone affect modulation, differentiation, and activation. Tysabri affects migration. Tecfidera affects depletion.

MS is now thought to be caused by the activation of T cells that migrate into the brain and start the disease process. Once in the brain, these T cells are locally reactivated by brain cells, and they recruit additional T cells and other immune cells called macrophages to establish the inflammatory focus in the brain that can be seen on MRI. Two types of T cells, CD8 and CD4, cause destruction of the myelin sheath and neurons resulting in neurological dysfunction over time.

It appears that B cells have a special function in triggering T cells that go to the brain. Expanded B cells are also found in the brains of patients with MS. When one targets B cells, one is affecting T cells. It does not appear that therapy against B cells works in MS by affecting antibodies, as was initially postulated.

Pulling back from this microscopic description, of course, are the lives and challenges of real patients.

I have seen and taken care of thousands of MS patients over the years, and certain cases stand out in my mind, often patients I have seen for many years—I usually know their family stories, their likes and dislikes, the nuances of their personalities. The patients who are the most prominent in my memory are the ones who represent the extreme ends of the MS spectrum—those who appeared to be cured and those I was unable to help. I think of an active young man with aggressive MS treated with cyclophosphamide at a time when he and his wife were starting a family. I now see him with children going to college. His disease shut off; he is on one of the new oral medicines and still exercising. The most painful are those on the other extreme, patients who have entered the progressive state of the disease despite our best efforts and use of strongest treatment. As the years have passed, I've watched them become severely disabled. I told one woman we would do everything possible to keep her out of a wheelchair. Twenty years later I saw her at the airport being wheeled onto the airplane. There's the former marathon runner who now hobbles with a cane. Every time I see a patient, I think about what the patient represents regarding the cause of the disease. I think about which questions I can ask, which experiments I can do to reveal secrets of MS that can potentially lead to a cure. We now have very effective therapy to treat relapses and inflammation in the brain. Now, when I see MS patients, the overriding question in my mind is whether they're developing the progressive form of the disease.

PROGRESSIVE MS

When MS shifts from relapsing to progressive, the crime scene changes. The processes are different, and the challenges are different. The crime scene in relapsing MS is that of infiltration of white blood cells and loss of myelin around the axons or nerve fibers. The crime scene in progressive MS consists of brain microglial and astrocyte activation and degeneration of brain axons. The origins of the crime scene in progressive MS can be found in the cells that enter the brain in relapsing MS, which then create a new crime scene that needs to be understood and confronted. Of course, one could argue, Don't let the cells enter in the first place and then you don't have to worry about the second crime scene. The concept of one crime scene setting up another occurs in most brain diseases. For example, in Alzheimer's disease, beta-amyloid accumulates in the brain, and the beta-amyloid triggers the accumulation of a protein called tau, creating a second crime scene.

Let's examine the crime scene in progressive MS. A very prominent feature is brain atrophy, or shrinkage of the brain. As we get older, all of us lose brain cells, and the brain slowly shrinks. Brain atrophy in MS patients

PROGRESSIVE MS

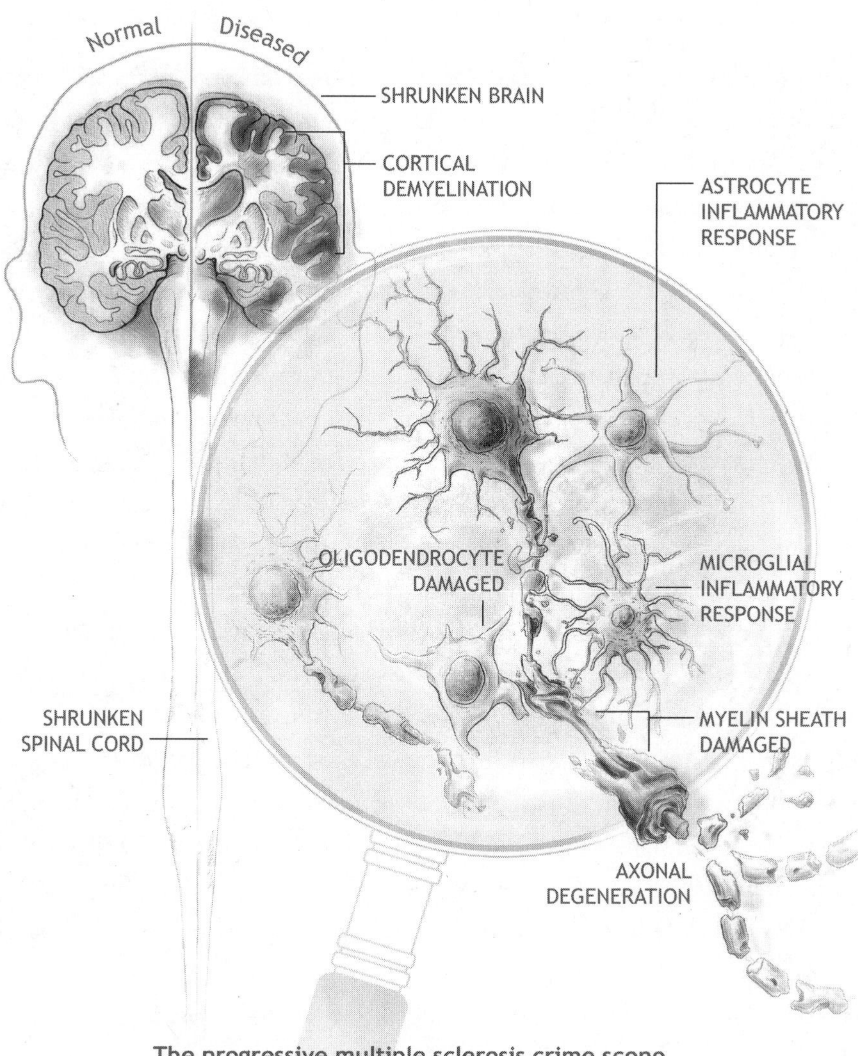

The progressive multiple sclerosis crime scene consists of a shrunken brain due to loss of axons, cortical demyelination, activation of astrocytes and microglial cells, and failure of remyelination.

occurs at a rate of approximately 0.5 percent to 1 percent per year, three to five times faster than in healthy brains. In MS, the atrophy is related to loss of nerve cells or axons, the threadlike part of the nerve cell that conducts impulses from cell to cell. Axonal degeneration begins early in MS. In his classic paper published in the *New England Journal of Medicine* in 1998, Bruce Trapp showed that axons are severed within MS lesions. A break in the axon causes a dead end of sorts for the nerve signal. Initially, the brain can compensate for these disruptions, but the more they occur, the harder it is to compensate. Trapp concluded that cutting the axons is linked to permanent neurological impairment.

Another feature of the crime scene in progressive MS is cortical demyelination. Cortical demyelination refers to the loss of myelin in the outer rim of the brain (the cortex), which impairs nerve conduction. Although cortical demyelination begins early in MS, it doesn't drive relapsing disease. The accumulation of cortical demyelination probably reaches a tipping point, after which it drives progressive MS.

A third feature is microglia and astrocyte activation. Microglial cells and astrocytes are local immune cells in the brain. They become activated and dysfunctional in MS and contribute to the progressive form of the disease. These cells may be at the center of much of the progressive degeneration in neurological diseases, including Alzheimer's and ALS. Microglia are the immune sentinels of the brain, but they can go astray and start to cause damage.

The fourth feature of the crime scene in progressive MS is failure of remyelination. The brain is able to repair itself to a degree by laying down new myelin where it has been damaged, but in MS there is a failure of remyelination over time. This leads to the accumulation of neurological deficits associated with progressive MS.

In textbooks MS is often categorized as a disease caused by an "inflammatory autoimmune process in the nervous system." This is true for relapsing disease and for the processes that initiate the disease, but it is not true for the progressive form. In progressive MS, the activated immune system becomes compartmentalized in the brain; this form is also driven by other processes that are no longer related to the immune system but are related to damage of the axons, which causes degeneration of nerve fibers, such as mitochondrial injury, oxidative stress, and glutamate toxicity. These are also processes that drive progressive neurological diseases such as Alzheimer's disease, ALS, and Parkinson's disease. This immune independence is one of the reasons that so many of the immune drugs that are successful at treating relapsing MS have failed to help progressive MS.

Clinically, there are two forms of progressive disease, primary and secondary progressive. Primary progressive starts with slow worsening with no clear attacks (flare-ups); whereas, secondary progressive evolves after initial attacks. For the most part, I believe this classification is artificial. The processes in both progressive forms of the disease are the same, and in many instances, primary progressive is simply a continuation of the relapsing form that was clinically silent. Independent of that, the approaches for treating progressive forms of the disease are the same. The FDA has further classified progressive forms of MS into active and non-active progressive MS. In active forms there may be occasional relapses and/or new MRI activity. Despite the best efforts of researchers around the world, successful treatments for progressive MS remain elusive for most patients.

Progressive MS has now come into focus for the entire MS community, and MS societies have created a progressive MS consortium that includes pharmaceutical sponsors. Close to fifty clinical trials of progressive forms of MS have been performed over the last thirty years, with over fifteen Phase 3 trials. Virtually all have failed. The advances in treating relapsing MS have not translated into successful treatment for progressive MS primarily because the crime scene and disease mechanisms in the two forms are different. Using drugs that target the immune system helps in relapsing disease but does not target the crime scene in the brain of progressive patients. Nonetheless, although most progressive MS trials have failed, hope blooms, as two drugs, Ocrevus and siponimod, have shown positive results in Phase 3 trials of progressive MS. Ocrevus, discussed above, has been FDA approved for the treatment of primary progressive MS.

Siponimod deserves special mention. Siponimod is an S1P-inhibitor, the same class of drugs as the oral drug Gilenya, which is FDA approved for treating relapsing MS. Gilenya and siponimod not only trap immune cells in the lymph nodes and prevent them from migrating to the brain, but they also cross the blood-brain barrier and reduce microglia and astrocyte activation, inhibiting neurodegeneration by acting on neurons and oligodendrocytes. This ability was expected to target the unique disease processes in progressive MS.

Because of this, between 2008 and 2011, Novartis conducted a Phase 3, randomized, double-blind, placebo-controlled trial called INFORMS in 970 primary progressive MS patients with Gilenya. I served on the scientific board of the trial and remember the excitement as we awaited the results. It would have been the first drug approved for progressive MS. Every FDA-certified trail must choose a primary clinical outcome or endpoint that must be reached for the drug to be approved. The trial of Gilenya had a novel clinical endpoint that combined the EDSS disability scale, a timed twenty-five-foot walk, and a nine-hole peg test measure of hand function. Imagine the disappointment when the trial failed to meet its endpoint.

As it turned out, Novartis had made a derivative of Gilenya called siponimod, which bound more selectively to S1P receptors on brain cells and bound less to S1P receptors on heart cells. After a Phase 2 trial showed siponimod was effective in reducing attacks in relapsing MS, Novartis decided to conduct a double-blind, randomized, placebo-controlled trial in 1,651 patients with secondary progressive MS—the largest trial of its kind—called EXPAND. This time an S1P inhibitor worked. The EXPAND trial reduced disability progression, which occurred in 32 percent of placebo patients and only 26 percent of siponimod patients. Although not dramatic, it was statistically significant and led to FDA approval. Remember, most trials in progressive MS failed with no significant findings in the primary endpoint. There were many secondary endpoints in the EXPAND trial, including MRI measures of the number of MS spots and of brain volume, all of which were achieved. Interestingly, worsening on the timed twenty-five-foot walk, a secondary outcome measure, was not affected by siponimod. Other disability measures besides walking may be more important in progressive MS, as we are exploring in our Progressive MS Initiative, discussed below.

A drug called ibudilast, which is currently being tested in progressive MS, also deserves special mention. Ibudilast is available in Asia for the treatment of asthma and post-stroke vertigo. It crosses the blood-brain barrier and can inhibit inflammatory immune molecules, which are increased in the spinal fluid of patients with progressive MS. In a Phase 2 trial of patients with relapsing MS, it didn't prevent new areas of MS from appearing on MRI but slowed brain atrophy and development of black holes on MRI, both measures of tissue destruction. Based on this, Robert Fox of the Cleveland Clinic led a Phase 2 trial of 225 progressive patients, funded by the NIH's Network for Excellence in Neuroscience Clinical Trials (NeuroNEXT). The primary outcome measure was rate of brain atrophy with secondary endpoints related to measures of tissue destruction, such as cortical atrophy and development of black holes. The trial was positive and showed a robust decrease in the rate of brain atrophy in ibudilast-treated patients. (It is interesting to note that MRI testing for ibudilast in progressive MS is different from the MRI metric for relapsing MS; gadolinium-enhancing lesions are classically used to screen for drugs for relapsing MS, as gadolinium lesions are linked to attacks.) The trial of ibudilast also had a positive effect on disability. Based on this, a Phase 3 trial in progressive MS is being considered.

Another drug that has shown promising Phase 2 results is called masitinib. Masitinib is given as a pill and targets microglia cells and mast cells, which are involved in allergic reactions. At the 2020 virtual ACTRIMS-ECTRIMS meeting, investigators reported that in Phase 2 trials sponsored by AB Science in Paris, masitinib slowed disability in patients with primary progressive MS and non-active secondary progressive MS when given over a two-year period in two 300-patient, placebo-controlled trials. A confirmatory study is planned. Masitinib is a protein kinase inhibitor that interferes with signaling pathways in the cell and decreases inflammatory processes. It appears to target microglial cells, which are a major component of the crime scene in progressive MS. Of note, masitinib is also being tested in Alzheimer's disease and ALS where microglia are felt to be important and where it has shown positive results. It is also being tested in asthma, certain cancers, and SARS-CoV-2 infection.

A new category of oral drugs called Bruton's tyrosine kinase (BTK) inhibitors is also being tested in progressive forms of MS by Sanofi and Genentech. BTK inhibitors appear to work by affecting both B cells and microglia. They have shown promising results in relapsing forms of MS, and Phase 3 trials are planned in both relapsing and progressive forms of MS.

A different approach for progressive MS has been the use of vitamins, such as high-dose biotin. The story of biotin begins serendipitously. When biotin was being tested for the treatment of a white matter brain disease called adrenal leukodystrophy, an MS patient was treated by mistake and seemed to respond to high-dose biotin. An open label trial showed similar effects, which were then further confirmed in a double-blind trial. It is not clear how high-dose biotin works, but it may enhance energy production and help dysfunctional axons and neurons function better. It only helps about a third of patients, and while results are not dramatic, I have had patients who report they are able to use their hands better or speak better with high-dose biotin obtained from compounding pharmacies. Unfortunately, a Phase 3, placebo-controlled FDA registered trial of biotin did not show a benefit. It is not clear why Phase 2 controlled trials worked, whereas the Phase 3 did not, but there was a much larger than expected placebo effect in the Phase 3 trial. Was I just observing a placebo effect in my patients? Dale McFarlin, an early pioneer in the study of MS, once said, "Put me in a trial and give me a placebo."

Other natural remedies include the antioxidant lipoic acid, which slowed brain atrophy in progressive MS in a small placebo-controlled trial.

Patients can buy high-dose biotin and lipoic acid over the counter, and many are taking it on their own.

One of my patients, who is doing quite well with her MS, nonetheless always begins her visit by asking, "I know I am doing well, Dr. Weiner, but when will we be able to remyelinate and repair everything in the brain? And what about stem cells?"

She is asking about what I term the second cure for MS, repairing a damaged nervous system. It certainly is possible to find compounds promoting remyelination in animals that work in the test tube, but translation to people is very difficult. One remyelinating strategy that has been translated into humans is being developed by Biogen. As it turns out, during development, the brain uses stop signals that prevent continual laying down of new nerve fiber tracks; one of these signals is called LINGO-1. Blocking the LINGO-1 stop signal resulted in remyelination in animal models. This approach was tested both in relapsing and progressive MS in the Biogen SYNERGY and AFFINITY trials using a monoclonal antibody against LINGO-1 called opicinumab. Unfortunately, these trials failed and Biogen has discontinued development of opicinumab for MS.

In another approach, Jonah Chan from the University of California San Francisco used an innovative screening assay to test FDA-approved drugs for their ability to stimulate the regrowth of myelin. He found that an over-the-counter antihistamine called Clemastine had myelin-repair properties; in a small Phase 2 clinical trial of MS patients who had optic nerve damage, he found modest improvement in transmission of electrical signals in the optic nerve. Many other approaches are being tried to identify remyelinating compounds in vitro, but again, translation to humans is difficult.

Finally, stem cell therapy is being tested in progressive MS. Stem cells are a powerful source of promise and hold out the magical hope of rebuilding or replacing damaged tissue. There are different types of stem cells—mesenchymal, hematopoietic, and pluripotent. In *mesenchymal* stem cell treatment, stem cells are isolated from a person's bone marrow, grown in the lab, and then reinjected into the blood or spinal fluid of patients. Mesenchymal stem cells are being tested in progressive MS and in ALS. *Hematopoietic* stem cell transplant, or bone marrow transplantation, involves treatment in which the whole immune system is destroyed with chemotherapy and bone marrow cells from the patient are reinjected. *Pluripotent* stem cells are transformed into primitive stem cells that can then be reprogrammed to become any cell type. The moral conundrum of using embryonic tissue to make pluripotent stem cells has been solved by the ability to make pluripotent stem cells from a person's own blood. Moreover, pluripotent stem cells are an important research tool in neurological diseases. At the Ann Romney Center, we are making pluripotent stem cells from MS and Alzheimer's patients to study the biology of the disease and as a screen for new therapies. Initial clinical trials of pluripotent stem cells have not been initiated in MS but have begun in Parkinson's disease and ALS.

In 2019, at the ECTRIMS international MS meeting in Stockholm, Dimitrios Karussis reported the results of a double-blind, placebo-controlled Phase 2 trial of mesenchymal stem cells conducted at Hadassah Hospital in Jerusalem in forty-eight patients with progressive MS. Mesenchymal stem cells were prepared from the patients' bone marrow and injected either into the blood or into the spinal fluid. Investigators found benefit on both clinical and MRI measures, especially in those who received the stem cells into the spinal cord. Further studies are now planned. A company called BrainStorm Cell Therapeutics is also testing mesenchymal stem cell transplants in four centers in the United States. In an open-label Phase 2 study they found improvement in measurements of neurologic function compared to a forty-eight, patient-matched cohort from our CLIMB study.

The path to effectively treat progressive MS is not a simple one, as the biologic processes created are complex and not easily addressed. However, many of the features of progressive neurological disease can be applied across diseases. For example, a trial of ibudilast that showed some promising results in progressive MS is now being undertaken in ALS, and drugs that affect microglia and astrocytes and/or energy metabolism could help more than one neurological disease.

PROGRESSIVE MS INITIATIVE

At the Ann Romney Center, we have created a comprehensive Progressive MS Initiative. The National MS Society has also made progressive MS one of its priorities. In our Progressive MS Initiative, we have identified several areas that we believe are at the front lines of effectively attacking progressive MS: imaging, biomarkers, animal models, microglial cells, new clinical measurements, new therapeutic initiatives, and the microbiome.

Imaging

The breakthrough in finding treatments for relapsing MS was the ability to measure inflammation in the brain via gadolinium enhancement seen on MRI. Gadolinium enhancement shows active inflammation. This allows us to test drugs on small numbers of patients. Successful treatments show a decrease in active MS spots, measured objectively with gadolinium enhancement.

In progressive MS, there is less gadolinium enhancement on MRI even as the neurological damage accumulates and brain atrophy continues. Different imaging approaches are needed for progressive MS, including measurements of grey matter atrophy. From pathological studies, we know that areas in the cortex (surface of the brain) play a crucial role in progressive MS. Until recently, we haven't been able to monitor those areas in living patients. In our Hale Building for Transformational Medicine, we now have one of the first 7-Tesla (7T) MRI magnets in the United States, which allows us to view the brain in greater detail than ever before. The 7T MRI is powerful enough to let us see and measure myelin loss in the cortex, something we haven't been able to do. We now need to track these patients over time using this powerful MRI to identify the features that are linked to progression of MS. This will allow us to better measure response to therapy and develop new treatments. The 7T MRI also allows us to see another aspect of progressive MS, inflammation on the covering of the brain (meninges). Inflammation of this part of the brain leads to an MRI sign known as leptomeningeal enhancement. As described earlier, in the CLIMB study, we have a cohort of extremely well-characterized MS patients who have been followed for years, including those who became progressive and those who did not. We can now delineate differences between these groups by using the 7T MRI to study the long-term relationship between 7T MRI and physical disability, cognitive impairment, and immune activation.

Blood and spinal fluid biomarkers

There aren't readily acceptable biomarkers for progressive MS, which has largely left diagnosis and treatment to clinical judgment. There is a new biomarker called neurofilament light chain or NfL, not to be confused with the National Football League. When the brain is damaged, neurofilaments are released from the damaged nerve fiber, just as cardiac enzymes are released from heart muscle when a person has a heart attack. Neurofilaments can be measured in the spinal fluid and in the blood. They are a general measure of brain damage and are elevated not only in MS but in other diseases as well, such as Alzheimer's disease, Parkinson's disease, ALS, and traumatic brain injury (TBI). They are not specific for progressive MS. We need a measure that tells us when a patient has transitioned from relapsing to progressive disease. We know the immune system changes when a person becomes progressive, and with sophisticated measurement of cells and metabolites in the blood and spinal fluid, biomarkers may be found to monitor progressive MS.

We found new biomarkers (microRNAs and antigen arrays) correlating with MRI and clinical features of progressive MS, and we're now exploring the use of complex techniques (single cell RNA-Seq and gene profiling of monocytes and the astrocyte marker GFAP.). Developing biomarkers for all diseases is needed, but doing so for progressive MS can be difficult, as the brain sits behind the blood-brain barrier.

Animal models

A third area necessary to understanding and treating progressive MS is a good animal model—the classic EAE animal model for relapsing MS is induced when a myelin protein is injected with an adjuvant, after which an animal has a single attack, and then recovers. We and others have developed an animal model for progressive disease in the mouse; after the mouse recovers completely from the first attack, it develops a slow, progressive disabling form of the disease. We have used this model to test a nasal spray that induces regulatory T cells to dampen activation of microglia and astrocytes.

Microglial cells

The major difference between progressive and relapsing MS is that in the relapsing stage, immune cells move from the blood to the brain. In progressive forms of MS, the immune response shifts to the brain itself where the inflammation is localized. The crucial immune cells residing in the brain and participating in localized inflammation are microglia and astrocytes. Animal and pathological studies have shown that these brain cells help drive progressive MS. In progressive MS, there is a loss of Dr. Jekyll (homeostatic) microglia and the appearance of Mr. Hyde (disease-promoting) microglia, which secrete pro-inflammatory chemicals that inflame astrocytes. Tarun Singhal in our Center has employed PET (positron emission tomography) tracers to image microglial cells in the brains of MS patients. We found changes that appear to be linked to progressive forms of the disease. This PET scan can be used to measure the effect of experimental treatments for progressive MS and other brain diseases such as Alzheimer's disease and ALS. We have also identified new surface markers on microglia and obtained complete genetic characterization of microglia and astrocytes both in animals and humans, opening an entirely new field of study not only for MS but for other brain diseases as well.

New clinical measurements

One of the major resources for understanding and treating progressive MS is the CLIMB study, which has enrolled over 2,500 patients. This longitudinal observational study began in 2000 when we established our MS Center and provides a unique opportunity to understand the course of MS in individual patients. Data from the CLIMB study have been used to identify new clinically based initiatives for progressive MS. Clinical scales for MS have traditionally been geared toward the relapsing form of the disease and have focused on ambulation. Progressive MS is more nuanced, with symptoms including fatigue, cognition, and impaired upper limb function, which may be a better measure of progression and response to therapy than lower limb function. Wearable technology now allows us to track movement and sleeping patterns continuously. In collaboration with Verily, a medical offshoot of Google, we found that biosensor data can correlate with MS disease severity. The massive CLIMB data set will also allow us to create a Progressive MS Risk and Prognosis Index. This index will also help us choose the best drug therapy for individual patients, and it will pinpoint patients whose treatment has resulted in NEDA (No Evidence of Disease Activity). Our CLIMB study is a unique resource to address the variables involved in NEDA. We hope to be able to predict the future course of the disease in an individual patient by matching the patient's profile with a similar profile found among all the patients in the CLIMB study.

New therapeutic initiatives

The primary goal of deciphering the mechanisms underlying progressive MS is to develop new therapies for progressive forms of the disease to have more shots on goal. As of this writing, a new therapeutic initiative for progressive MS is ready to be launched at our Center. This trial is one in which the monoclonal antibody anti-CD3 is given by a nasal spray. In an animal model of progressive MS, we discovered that nasal anti-CD3 slows disease progression, decreases brain damage to nerve cells, and dampens inflammation in microglial cells and astrocytes. It took five years of work in the laboratory before the positive results were published. Nasal anti-CD3 works by interacting with the lining of the nasal cavity and then stimulating regulatory T cells in the cervical lymph nodes in the neck. These regulatory T cells travel to the brain, where they secrete the anti-inflammatory chemical IL-10, which in turn dampens the local inflammation in the brain. Although it is given nasally, it does not work by directly entering the brain, as is the case for other nasally delivered drugs, such as nasal insulin, which is being tested in Alzheimer's disease. We successfully completed a Phase 1 dosing trial in healthy individuals in which a nasal anti-CD3 monoclonal antibody called foralumab from Tiziana Life Sciences was found to be safe and to modulate the immune system. We are now ready to test nasal foralumab in patients with progressive MS.

A fascinating offshoot of our initiatives in progressive MS is that the approach applies to the other progressive neurological diseases we are studying, diseases that also have inflammation of microglia and astrocytes, such as ALS and Alzheimer's disease. We are planning trials of nasal anti-CD3 in Alzheimer's disease, and, in collaboration with Merit Cudkowicz at Massachusetts General Hospital, we are considering a trial of nasal anti-CD3 in patients with ALS.

Interestingly, we hypothesized that nasal anti-CD3 might be of benefit to those suffering from COVID-19 by dampening the overactive immune response that occurs after viral infection. We carried out a small pilot trial in Brazil and indeed found that nasal foralumab dampened inflammation and helped clear lung abnormalities.

THE MICROBIOME

The gut microbiome is a major new frontier in medicine. A revolution has occurred with the recognition of the key role that the bacteria in our gut play in the biology and mechanism of disease. These gut bacteria are known collectively as the microbiome. This new area is rapidly developing in medicine and has the potential to help not only MS but other neurological diseases, too. Research is showing the microbiome affects not only our ability to process food but also our immune system and how well we fight off disease. Additionally, it has become clear that there is a gut-brain axis, in which changes in the gut can affect the brain and vice versa. This is true not only for MS but also for all the diseases discussed in this book, as well as other conditions, including autism, epilepsy, and obesity.

I always learn from my patients. Whenever I see one, I consider what they tell me in the context of the cause and treatment of MS. I often consider if their symptoms are related to the disease itself or are a reaction to having the disease. I remember one such moment when a patient of mine, a middle-aged woman on a scooter, asked about antibiotics. She was an avid Red Sox fan and came to the clinic wearing a Red Sox jersey and cap. She told me she felt better when she took antibiotics and wanted me to prescribe them for her. I asked her why she thought she improved and she said she didn't know. It was just something she felt. When I hear a story like this, I always ask, Is this simply the placebo effect, or is there a biologic basis for it? In her case, there was potentially a real cause and effect of the antibiotics. The bacteria in her gut could be altering her microbiome although we didn't know how. Antibiotics drastically affect the microbiome as they kill many of the bacteria in the gut, both good and bad. In fact, in the laboratory, to study the role of the microbiome in animals, we treat them with antibiotics to remove the bacteria from the gut. This can also be done by studying animals raised in a germ-free environment.

When I speak to patients about the microbiome, I begin by asking them how many bacteria they think there are in our gut. The answer usually comes back millions, then billions, then hundreds of billions. I tell them it's more. In fact, there are trillions of bacteria in our gut, and bacteria cells equal human cells in our body. Most of the bacteria in our body are in our gut. By studying and understanding the different species of bacteria in our bodies and how they relate to our health, we can one day manipulate them as a form of treatment.

The microbiome is such a new area of exploration that it wasn't until relatively recently that we could identify these bacteria. In December 2007, the National Institutes of Health (NIH) launched the Human Microbiome Project. At the time, the NIH realized how little we understood about the trillions of bacteria in our bodies. The goal was to identify the microorganisms in both healthy people and those with disease. This was done by extracting bacterial genetic material from the gut. In 2012, the Human Microbiome Project Consortium achieved a major milestone. The group published the structure, function, and diversity of the healthy human microbiome. It was a road map for the study of the microbiome.

We evolved while living in a dirty world without sanitation and with exposure to all of the disease-producing bugs and beneficial bacteria in the environment. We now realize that exposure to microbes is not necessarily bad. In recent years, the hygiene hypothesis has been proposed to explain the rise of certain diseases. The hypothesis states that the immune system is conditioned by exposure to infectious agents in the environment, and if we are not exposed to infectious agents because we no longer live in a dirty environment, the immune system is unable to deal with disease, making certain diseases prevalent. This may especially be true for autoimmune diseases such as MS and juvenile diabetes.

Our exposure to microbes in the environment actually begins at birth. When a baby is born, the baby is colonized by the many bacteria in the vaginal canal. This colonization doesn't occur after caesarean section, and children born by caesarean section have a greater incidence of allergy and asthma when they grow up. Other intriguing studies suggest that there is less asthma in homes with more children and a dog. However, if your dog stays inside, you have a greater chance of developing allergies when you grow up than if your dog goes outside.

In addition to environment, other factors affect the gut microbiome in ways that could trigger disease. We know, for example, a more Westernized diet has coincided with a higher incidence of MS in Japan. I was recently in Japan, where I met with MS scientists in Tokyo and gave a lecture on the past and future of mucosal tolerance. I also spoke about our work on nasal spray for progressive MS, along with a report on our use of probiotics to treat MS.

One morning, at the National Institute of Neuroscience and Multiple Sclerosis Center, Takashi Yamamura held a scientific session for me in which we discussed their ongoing research. One slide caught my attention: it showed the incidence of MS in Japan over the past two decades. Where MS was uncommon twenty years ago, it was now more and more common. Why was there a dramatic increase of MS in Japan? We knew the genetics hadn't changed, so it had to be something in the environment. Other environmental risk factors for MS hadn't changed in Japan, including delayed Epstein-Barr virus infections, UV exposure, or salt intake. The number of smokers went down. Everyone thought the change had to be due to the microbiome and what people were eating.

Later that month, at the 14th Annual International Neuroimmunology Meeting in Brisbane, Australia, Takashi Yamamura showed a slide of a typical Japanese breakfast with fish, rice, vegetables, and miso soup with the title "Westernization and Change of Eating Habits: Traditional Japanese Breakfast Might Prevent the Occurrence of MS." Over the years there has been an increased intake of milk and dairy products in Japan and a decreased intake of dietary fiber. In the past, patients Yamamura had diagnosed with MS in Japan had all lived abroad for a period of time before their onset of MS. Although anecdotal, he then told the story of a well-known CEO of an IT company in Japan whose first signs of MS occurred in 2006 followed by serious relapses. His MS dramatically improved when he retired, started working as a farmer, and only ate homemade beans, Miso soup, and vegetables. Here was an experiment that appeared to provide evidence that affecting the microbiome might help MS. However, as with any anecdotal observation, proof would require isolating bacteria that changed with the new diet and showing that giving them to MS patients in a controlled trial helped MS.

Is the microbiome indeed abnormal in MS? This question was first investigated in the EAE animal model of MS in a strain of mice that spontaneously develops the disease. Hartmut Wekerle's group in Germany took these animals and raised them in a germ-free environment so they had no microbiome. These animals did not develop MS. However, if the researchers recolonized the gut of these mice with bacteria a few weeks after birth, they developed MS. This shows the microbiome can be crucial in triggering MS in a genetically susceptible animal. The big question, of course, is what is it in the microbiome that triggers disease in these animals? Studies in EAE have since shown that a type of bacteria called segmented-filamentous bacteria, which is invasive and sticks to cells in the gut, can drive the expansion of autoimmune disease–inducing TH17 cells in the gut. Colonizing the gut of mice with these bacteria results in more TH17 cells in the spinal cord and more disease.

However, nothing is ever that simple. Remember, the body is built on regulation, with opposing forces balancing each other to maintain equilibrium. It turns out you can trigger both regulatory T cells *and* autoimmune disease–inducing TH17 cells in the gut. Lloyd Kasper's group at Dartmouth's Geisel School of Medicine treated animals with broad-spectrum oral antibiotics to reduce the gut microflora. The animals did not develop more EAE; they developed less. This change wasn't seen in animals in which the antibiotics were given intravenously; it was a direct effect of the antibiotics on the gut. Kasper identified regulatory cells in the gut that appeared to be induced by the antibiotic treatment, and he was able to transfer protection from animals treated orally with antibiotics to other animals. Others have found that treating mice with components from a beneficial bacteria such as *Bacteroides fragilis* expanded the numbers of intestinal regulatory cells and protected mice against disease. We have done similar experiments in our laboratory looking to identify those bacteria in the gut that promote EAE and those that suppress it.

These mouse studies raise the big question of whether or not the microbiome in MS patients is normal or abnormal. And if it is abnormal, how is it abnormal? Many people believe that if we better understand the gut and modulate it early in life, we may be able to prevent MS.

I have had a long-term interest in the gut and the potential to induce regulatory cells in the gut. One of my most important scientific experiments was to successfully treat the EAE animal model of MS by feeding proteins isolated from the myelin sheath to induce brain-specific regulatory cells. The concept is called oral tolerance, and I spent over a decade developing this idea. We published many papers, started a company, and carried out large Phase 3 trials not only in MS, but also in arthritis and diabetes, only to have them fail. I never knew exactly why those trials didn't work, but as the understanding and focus on the gut microbiome moved front and center in biologic investigation, I wondered whether the microbiome was different in MS patients. Could there be a defect in the microbiome that explained why my oral tolerance experiments didn't work in people?

To study the microbiome in MS, it was necessary to collect stool samples from our MS patients. It wasn't easy to explain to the patients, and people who worked on our CLIMB study called the microbiome study the "poop study." I realized that rather than asking patients whether I could have a sample of their poop, I could lead with a different question—asking if they would help us try to develop a vaccine for MS. People said yes, of course, and then I explained that in order to do that, we needed to study stool samples.

It was difficult to get funding for our study. Our grant request to the National Multiple Sclerosis Society had to go through two rounds of review, as reviewers said there was no preliminary data to show that the gut was abnormal in MS. This, of course, is the catch-22 of scientific research. Money is given when one already has results but rarely when there are no preliminary findings. This stumbling block is changing with pilot grants and new grants from the NIH called R21 grants that don't require preliminary data. Nonetheless, even for these grants, the general rule remains: good preliminary data drive research funding. Put another way, it's hard to get money to drill an oil well, but everyone will give you money once you've found the oil. Dennis Selkoe and I realized this when we started our Center for Neurologic Diseases in 1985. Philanthropic start-up funds have made all the difference in our ability to drill for oil.

I strongly believe that all researchers need flexible exploratory funds to chase ideas. These funds are hard to come by through normal granting mechanisms and therefore may only be possible through philanthropy. So, using philanthropic funds, we began collecting poop from our MS patients. We collected samples from sixty MS patients and forty-three from healthy controls. This was not an easy study. There were many factors that had to be taken into account—we couldn't accept patients with gastrointestinal disease or recent use of antibiotics. All the study subjects had to complete a detailed dietary questionnaire.

The gut can be considered a giant ecosystem, and we were looking for differences between the MS patients and the healthy controls. For example, if an ecosystem consisted of all the trees, animals, flowers, and bugs, was there equal representation of those species in the gut of MS patients versus the healthy controls?

There are two ways to measure the ecosystem of the microbiome, known as alpha and beta diversity. Alpha diversity measures the number and evenness of the distribution of bacteria in the microbiome of a single population. In other words, how many different types of fish are there in the Pacific Ocean? Beta diversity measures the overlap of bacteria between two populations. In other words, how many types of fish are found in both the Pacific and Atlantic Oceans? In our study, the alpha and beta diversity were not dramatically different between MS patients and healthy controls. However, as we dug deeper into the data, we began to find many interesting differences.

Our major findings centered on three bacteria and offered tantalizing clues about how the microbiome not only affected MS but also could be used as a diagnostic tool. When I trained to be a neurologist, I had to learn the names of tracts in the brain. Now I was confronted with learning the names of bugs in the gut. Patients with MS had increased amounts of microbes called *Methanobrevibacter* and *Akkermansia* and decreased amounts of a bug called *Butyricimonas*. We postulated that *Methanobrevibacter* and *Akkermansia* drove inflammatory responses, whereas *Butyricimonas* drove regulatory type T cells. Surprisingly, we later found out that elevated *Akkermansia* was actually a beneficial compensatory response in the gut and that treating mice with *Akkermansia* promoted regulatory T cells and ameliorated disease. We also measured the effect of MS treatment on the microbiome and found that some species increased with treatment, such as *Prevotella* and *Sutterella*.

We then wanted to see whether these organisms affected the immune system. We carried out sophisticated assays of immune cells in the blood of MS patients and found that genes involved in pro-inflammatory responses were linked to the bacteria we found in the gut. In a fascinating offshoot, when we read about the *Methanobrevibacter* we had found in the microbiome, we learned these bacteria were methane-producing organisms. This led us to do breath tests on another cohort of MS patients, and we found indeed that there was increased methane in the breath of some of our MS patients.

With these results, it was no longer difficult to get funding. We had struck oil. Furthermore, we were no longer the only group beginning to study the microbiome in MS, and many other publications began to appear. These studies also showed differences between the gut microbiome of MS patients and healthy controls. Some of the results were consistent with ours; others were different.

Although cautious, we concluded our article as follows: "It is possible that treatment strategies of MS in the future may include therapeutic interventions designed to affect the microbiome, such as probiotics, fecal transplantation and delivery of constituents of organisms isolated from the microbiome. In addition, characterization of the gut microbiome in MS may provide biomarkers for assessing disease activity and could theoretically be an avenue to prevent MS in young at risk populations."

The field is now growing rapidly. In addition to our own studies, we are part of an International MS Microbiome Consortium established by Steve Hauser and Sergio Baranzini at the University of California San Francisco, again driven by a large philanthropic gift. Countless more studies will be undertaken, and many more papers written before we know the full role of the microbiome in MS. And—as is the case in all of science—some of what we've found will not turn out to be valid, while other findings will be strengthened. In any event, a role for the microbiome in MS will only become clearer. As we will see in the next chapters, the microbiome plays a role in each of the other neurological diseases discussed in this book. In fact, the microbiome touches all aspects of medicine: for example, the response to cancer drugs is linked to the microbiome.

Three papers demonstrate how much we are learning about the role of the microbiome in MS and where the field is heading. One comes from Hartmut Wekerle's group, which showed for the first time that gut bacteria play an active role in T cell activation and could play a part in triggering MS in humans. The researchers arrived at their groundbreaking conclusions by looking at thirty-four human monozygotic twin pairs that were discordant for MS. These were genetically identical twins, recruited from all over Germany, half with MS and half without. When researchers checked the twins' gut bacteria, they found a difference. Similar to our research, they found increased *Akkermansia* in the MS twins. Even more significantly, when they injected the microbiota from both the MS and healthy twins into genetically susceptible mice raised in sterile environments, the gut bacteria from the MS twins induced a higher incidence of autoimmunity than the healthy twins' gut bacteria, and almost all of these mice developed MS-like brain lesions. Wekerle's group also found the bacterium *Sutterella* had a protective effect. The results raise the possibility that both protective and disease-causing bacteria can be found in the human microbiome.

A second study, by Sergio Baranzini at UCSF, looked at differences in the gut bacteria between seventy-one MS patients not undergoing treatment and seventy-one healthy controls. Baranzini's group found two bacteria significantly associated with MS: *Akkermansia* and *Acinetobacter*. When they transplanted microbiota from MS patients into germ-free mice, they found the mice had more severe symptoms of disease and fewer IL-10 producing cells. IL-10, also called interleukin 10, is an anti-inflammatory chemical.

Another group, led by Iliara Cosorich at the San Raffaele Scientific Institute in Milan, Italy, measured the frequency of intestinal TH17 cells. Analyzing the microbiota from small intestinal tissue, the researchers found that MS patients with high disease activity and increased intestinal TH17 cell frequency showed a higher *Firmicutes/Bacteroidetes* ratio, more *Streptococcus*, and decreased *Prevotella* strains, compared to both healthy controls and MS patients with no disease activity. Consistent with this, others have found that *Prevotella* can protect mice from severe EAE.

Although these three studies clearly identify abnormalities of the microbiome in MS patients, the mechanism by which the microbiome relates to MS is not simple and has many facets. First, it is theoretically possible that the gut holds the key to the origins of MS: although people have not been able to find an MS virus or MS infectious agent, it is possible that the infectious agent triggering MS is found in the gut. This agent likely has structures on its surface that are similar to brain structures, and because of this similarity, the agent triggers cells that go to the brain and cause disease, just as a strep throat can lead to rheumatic heart disease because the *Streptococcus* bacteria has structures on it that are identical to structures on heart valves. The immune system attacks the brain because it thinks it is attacking the bacteria. Thus, brain cross-reactivity with a specific organism may trigger MS. If that is the case, vaccination strategies to prevent infection by the triggering organism become obvious.

Second, it may be that there is not just one bacterium that triggers MS, but the overall composition of the microbiome creates susceptibility to MS. Bacteria could provide a one-two punch in the gut to trigger MS—the first being cross-reactivity with a brain protein and the second being modulation of the immune system toward an inflammatory response. Use of antibiotics, environmental exposure, and diet could each trigger MS in a genetically susceptible person. This appears to be what is happening in Japan with the prevalence of Western fare.

Third, the gut could promote MS by modulating the systemic immune response, driving production of pro-inflammatory TH17 cells versus regulatory T cells. Fourth, the drugs we use for MS may act in part through the gut, and we know that certain drugs used to treat cancer are dependent on the gut microbiome. Fifth, there is a direct relationship between the gut and the brain. On a simplistic level, this is obvious. There are hormones secreted by the brain that affect the gut. When we are nervous or frightened, we feel it in our gut. Also, nerve fibers from the brain go to the gut and vice versa. Sixth, microbes can talk to our brain by producing neurotransmitters and other chemicals that can affect brain cells. Animals that are raised in a germ-free environment or are treated with antibiotics have altered behavior and less developed microglial cells in the brain. Others have found that treating mice with bacteria that make the inhibitory neurotransmitter GABA have less depressive behavior. We have identified specific microbes in patients with progressive MS that are linked to depression and fatigue.

We are moving toward treating MS by modulating the microbiome. We have conducted a trial of probiotics in patients with MS and found that probiotics reduced pro-inflammatory signals in the blood. In animal studies, the probiotic we used in MS was shown to decrease damage in animals with spinal cord injury.

Research scientists are scrambling to find specific bacteria that could be administered to MS patients and have a beneficial effect on the disease. My long-term interest in modulating the gut via oral tolerance has been rekindled, and we are stimulating the gut and nasal immune system with anti-CD3 monoclonal antibody. However, it is not necessary to know the specific bacteria involved to treat via the gut. A burgeoning area in medicine is the use of fecal microbiome transplantation. If it sounds strange to ask people to give samples of poop, it is even stranger to contemplate transferring poop from one individual to another. However, this is indeed happening, and it is now accepted medical practice for patients suffering from a severe infection of the colon caused by *Clostridioides difficile*, also known as *C. diff.* We know human colonic microbiota provides resistance against various bacterial pathogens, and it is considered to be a key determinant in the prevention of *C. diff.*

Antibiotics can affect the microbiome diversity in the colon, resulting in loss of resistance to infection. Trials of fecal microbiome transplant (FMT) have shown dramatic results in *C. diff.* Which components of the microbiome provide resistance are not known, but the rectal transplantation of feces from a healthy pre-tested donor combined with stopping all antibiotic use can successfully treat more than 90 percent of patients with recurrent *C. diff.* At the present time, fecal microbiome transplant is being used widely in *C. diff* infections and does not require specific FDA permission, although trials using FMT in other diseases require FDA permission. FMT is currently being considered for the treatment of many diseases, and a nonprofit organization called OpenBiome is collecting feces to be used for treatment trials.

There are initial reports that transfer of fecal microbiota from humans who are thin increased insulin sensitivity when transferred to humans with metabolic syndrome. Another open label study suggests that microbiota transfer may improve autism symptoms. Research is in its infancy, and we still do not know how to specifically manipulate the microbiome and which bacteria to transfer. Nonetheless, centers, including ours, are considering fecal microbiome transplants in patients with MS.

Although it is now clear that the gut has an important role in MS, its exact role and how to manipulate it safely have yet to be determined. Nonetheless, treatment with MS-specific probiotics is likely in the future. Other approaches may target detrimental gut bacteria in MS. We have found that small RNAs in the gut of MS patients can modulate bacteria and ameliorate the EAE animal model of MS. Understanding the microbiome is a new frontier in our understanding and the treatment, not only of MS but, as we shall see, of other neurological diseases as well. Because of this, we recruited a microbiologist, Laurie Cox, to be a full-time faculty member at the Ann Romney Center for Neurologic Diseases. Henry McFarland, a pioneer in the study of MS at the NIH, now emeritus, found himself in retirement organizing a conference on the role of the microbiome in neurological diseases, something he never imagined doing in his years at the NIH.

Many patients ask me what I recommend regarding the gut. Is there a specific diet? Should one take probiotics? Presently, there is no clear answer. At our MS Center, we recommend a healthy diet, and if one wants a specific diet, a Mediterranean-type diet is preferable. There are many books on diet and MS. One dietary factor is vitamin D. MS is worse in patients who do not have adequate blood levels of vitamin D, so we monitor this religiously in all our patients. There are also diets such as the Wahls Protocol and the Swank diet. The Wahls Protocol is named for Terry Wahls, a doctor who began studying food and vitamins after being diagnosed with MS. Her diet is a version of the Paleo diet, which avoids wheat and processed foods and is designed to mimic the way our prehistoric ancestors ate. The diet reportedly helped Terry Wahls move from a wheelchair to climbing mountains. The late Roy Swank trained at the Brigham and Women's Hospital and advocated a very low-fat diet for many years, work now carried on by the Swank MS Foundation. Clearly, the purported success of these diets needs to be taken with caution. It will be many years before we fully understand how the gut affects MS and other diseases of the brain, but it is clear that understanding and treating MS will in the future involve understanding the microbiome.

SHAM "CURES"

One component of carrying out research on diseases such as MS is not taught in medical school or the laboratory: dealing with and confronting treatment claims and approaches that do not pass scientific scrutiny but offer magical cures and create false hope. Many of them abound in medicine, and they can be disruptive.

On a Sunday night in December 1990, I turned on the television to watch the news. As I began to flip through the channels, I came across *60 Minutes*, and the words "multiple sclerosis" grabbed my attention. An attractive woman with a diagnosis of MS was describing how she improved after having her teeth fillings removed.

"And how soon after you had the fillings removed did you feel better?" the reporter Morley Safer asked.

"The next day," the woman said. "And after that I no longer had to use my cane."

"And how have you been since then?"

"I feel better than I've ever felt," the woman said.

I cringed. "Oh my God," I said to myself. "Why in the world did they put that on *60 Minutes*?" The newsmagazine program was one of the most watched shows on television, with a reputation for impeccable journalism. The scientific community had long dismissed the notion that replacing silver fillings (which contain mercury) could somehow help MS. The treatment was totally unproven, with no evidence to back the hypothesis. If fillings are related to MS, why do millions of people who have fillings not get MS? And there's more mercury in the food we eat, especially fish, than in our fillings. Nonetheless, how does one explain her response to her fillings being removed?

MS can have naturally occurring remissions, and it's possible that this is what happened with her. Patients may also unconsciously exaggerate or embellish their symptoms, which could disappear quickly. That night my phone began to ring. A patient from New York called. "Did you see the show? Should I have my fillings removed?" I told her that this is not something she should pursue. Then she asked, "But if it isn't dangerous, how can it hurt?" That question sits at the heart of treatments for MS that have no scientific basis in fact. Why not try something if it has no side effects? Furthermore, if medical science can't come up with a cure for the disease, perhaps scientists are missing something.

Three days after the segment appeared, the National MS Society wrote to Don Hewitt, the executive producer of *60 Minutes*, explaining there is no validity behind the story. When the show was aired in 1990, MS patients were particularly powerless in the face of their disease. The FDA had yet to approve a drug for MS. In fact, before there were treatments for MS, the National MS Society published a book called *Therapeutic Claims*. The word *claims* implied that these were things people claimed but were not necessarily true. Now that we have effective treatments, the National MS Society no longer has a book called *Therapeutic Claims* but instead a book on therapeutic options.

There have been many false treatments for MS over the years, including snake venom, bee stings, and hyperbaric oxygen. An article on hyperbaric oxygen was published in the *New England Journal of Medicine*, in 1983, in the same issue that we published our findings with cyclophosphamide. Hyperbaric oxygen is pure oxygen delivered in a chamber where atmospheric pressure is increased. Using hyperbaric oxygen to treat MS made no sense, and subsequent studies showed the therapy had no benefit. It is not clear how the researchers obtained the data published in the article; most likely there were either flaws or perhaps even manipulation of the data. Unfortunately, that can and still does happen in medicine, as we've seen with a recent fraudulent study on new ways to make stem cells, published in *Nature*.

Now that we have FDA-approved treatments for MS, you would think episodes like the 1990 segment on tooth fillings wouldn't happen again. This, however, is not the case. In 2009, an Italian physician named Paolo Zamboni published a study of something he labeled chronic cerebrospinal venous insufficiency, or CCSVI, for the treatment of MS. On October 14, 2009, Zamboni visited our MS center and presented his work on CCSVI. He had just published a paper in the *Journal of Neurology, Neurosurgery, and Psychiatry* in which he studied sixty-five patients with MS and compared them to 235 controls that included healthy subjects and patients with other neurological diseases. He reported that the patients blindly underwent transcranial and extracranial high-resolution Doppler ultrasound to measure venous outflow. Remarkably, he found venous outflow abnormalities in every one of his MS patients but not in any of the controls. A study like this with perfect results is hardly ever seen. He also found different patterns in relapsing and secondary progressive MS as compared to primary progressive MS.

When I spoke to him, I asked how his findings fit with the known biology of MS that centered on the immune system and proof that drugs helped MS by stopping cells from entering the brain. He had no real answer, only to say that what he'd found was real. No one expected that CCSVI would explode into a major issue in multiple sclerosis, triggered by Canadian media reports.

The Canadian television piece began with a breathless introduction that went as follows: "This week, a stunning medical treatment—a revolutionary treatment for a most debilitating disease. It is multiple sclerosis, and Canadians are especially hard hit. Our team has an exclusive first look at a radical new approach that suggests the disease is not just an autoimmune disorder as has long been thought. A relatively simple procedure called 'the Liberation Treatment' could free MS patients from a lifetime of suffering. A scientific breakthrough, but one inspired by love."

The journalist, Avis Favaro, filed her piece for the Canadian Television Network from Italy, where she reported on Paolo Zamboni's "amazing theory that could turn the diagnosis and treatment of MS upside down." She told of Zamboni's wife having MS and his wanting to help her. Zamboni himself suffered from a neurological disease that left him partially disabled. Favaro showed disabled MS patients hoping for a treatment, and she reported that no one had ever conclusively proven what starts MS (although technically true, we have very good theories about what starts MS).

In the TV piece, Zamboni reported that unusually high levels of iron were found in the brains of MS patients and said, "This is very important news, because iron is very dangerous." His studies found impaired venous drainage involving both the jugular veins and deeper veins in the body. He stated that every single patient had some sort of vein narrowing. Favaro egged him on. "Did you say, 'This is it'? Did you jump up and down? Were you excited?" Zamboni responded that he was absolutely excited, but that when he spoke to other neurologists, they were not interested enough to begin treatment. He did, however, find a neurologist who liked the idea and began sending patients to Zamboni for testing. The neurologist appeared in the piece and added, "The images of narrowed or blocked drains called strictures were irrefutable."

The program then featured a man called Augusto Zeppi, who'd been suffering from regular MS attacks for nine years and had fatigue and trouble walking. On camera, he said, "Everything I was dreaming for my future adult life would never be possible. Game over." He was told that his jugular veins were blocked, and he became the first to undergo the treatment, a vascular shunting called percutaneous transluminal angioplasty. Zeppi said the treatment restored his health. He was shown drinking coffee in a coffee house. "I forgot what it's like to have MS. I feel well. I feel like I did fifteen years ago. There was a time I could not play tennis with my son. Now I can," he said tearfully. Favaro ended her piece with, "You can play tennis with your son now. He gave you back your life."

Many times, I have sat with a disabled MS patient who pleaded, "Can't you do something to help?" I tell them that I wish I could just pull something out of the drawer in my office and give it to them, but I can't. We still don't have treatments to reverse established disability, but we have many patients with relapsing MS who no longer have attacks and tell me they don't feel like they have MS.

The neurologist who sent Zamboni his first patient for testing said he liked to think outside the box. "Thinking outside the box" is a commonly used phrase. It implies that our theories and understanding are somehow myopic and wrong, and the only way we will solve the problem is to explore nontraditional ideas. I've always been troubled by this phrase, because in science the box we are in is created by hard biologic facts and building blocks. At some scientific meetings when people say we need "out-of-the-box" thinking, I often say we need to think creatively to understand what's already in the box that we aren't yet exploring.

CCSVI began to gain traction. Patients felt there was a long-lost cure that was now available for MS. The fact that the treatment was called the Liberation Treatment made it only more exciting, as patients understandably wanted to be liberated from their disease. Zamboni also referred to his treatment as the "Big Idea."

The medical community did not embrace the theory. It made little sense, and how could every MS patient, without exception, have this condition? Nonetheless, patients flocked to clinics both in the United States and overseas to have vascular shunting done, in which stents were placed in the jugular vein to open what was thought to be a narrowing there. There were anecdotal reports that some people were helped, but there were also complications and at least two deaths were linked to the experimental procedure. MS shunting was halted at Stanford in 2010 after a fifty-one-year-old patient died from a fatal hemorrhage while on a blood thinner following the procedure. The same year, a thirty-five-year-old Canadian died after undergoing the procedure.

Some physicians wondered about Zamboni's motives. They raised the question of his financial ties to one of the machines sold to measure the vascular insufficiency. Nonetheless, Zamboni's theory had proponents. There were some doctors who argued that if we could prove CCSVI was the underlying cause of MS, it would change the face of MS. The cat was now out of the bag, and special interest groups began advocating for the testing of CCSVI. Proponents argued that scientists were in the back pocket of Big Pharma and were hiding the cure from the patients.

Medical science is evidence based and requires rigor of design and, most important, reproducibility. For the public and patients, however, often the most powerful evidence is anecdotal. Journalists love anecdotes, particularly heart-wrenching stories of the sick or the underdog. Health Canada was presented with a petition of almost twenty thousand signatures from people pleading for something to be done. An expert panel was convened that recommended careful monitoring of patients. Some $5.4 million was made available for testing, diverting funds from other MS research.

Four years after the Zamboni article was first published, CCSVI was finally put to rest. A study called CoSMo evaluated almost two thousand individuals in Italy and found no evidence for CCSVI. A prospective randomized trial called PREMiSe showed no effect. Articles began to appear, such as "What Went Wrong? The Flawed Concept of Cerebrospinal Venous Insufficiency" and "CCSVI Deconstructed and Discarded."

In a 2014 article entitled "Venous Angioplasty for CCSVI in Multiple Sclerosis: Ending a Therapeutic Misadventure," Dennis Bourdette and Jeff Cohen describe an experiment in 1889 by Charcot, the scientific giant who was one of the first to identify MS. Charcot was a proponent of "suspension therapy" for types of syphilis in which patients were suspended from a harness attached to their chin and the back of the head for several minutes. The goal was to stretch the spinal cord to improve circulation. Although some patients' symptoms seemed to improve, there were side effects, including death by strangulation. Despite his initial enthusiasm, Charcot realized that it made no sense, and he abandoned suspension therapy. Like CCSVI, suspension therapy was a treatment founded on faulty theory.

In today's age, the world of social media creates an alternative universe that can drive medical hypotheses, independent of science. Nonetheless, in late 2017, Zamboni himself published an article in *JAMA Neurology* concluding that percutaneous transluminal angioplasty was "a safe but largely ineffective technique," and CCSVI was finally put to rest.

It was a wonderful story. Sadly, it was not true. It was indeed a "therapeutic misadventure."

Many would argue that we cannot know for sure when an anecdote is real or false. In some instances, anecdotes are tested and turn out to be real. This appears to have been the case with the psoriasis drug Fumaderm, taken by a neurologist in Germany with MS who appeared to improve. On the other side of the equation, even studies that appear to be real may sometimes wind up being false. A major pharmaceutical company paid billions of dollars to women who sued after their silicone breast implants appeared to cause autoimmune disease, only to learn later the studies were flawed.

CURING MS

How then will we ultimately cure MS? What will be the real "Liberation Treatment," and what is the real "Big Idea"? I believe we have made remarkable progress in our understanding of MS and are effectively curing many people. Our biggest challenge now is for progressive MS—how to identify the pathways that drive it, how to measure it, and how to prevent it. It's said Einstein figured out the theory of relativity by conducting a thought experiment while riding on a trolley in Bern, imagining what it would be like if he was traveling on a beam of light. If I carry out such a thought experiment in MS, it leads me to conclude that MS will ultimately be cured in the same way as polio. MS is a disease of the immune system, and because the immune system responds to the environment, MS must be triggered by environmental factors in a genetically susceptible host.

I imagine a time when children or people at risk will be given a vaccine, most likely targeting the gut immune system. The vaccine will prevent the immune system from triggering the disease, just as vaccines prevent polio and smallpox. I envision an MS vaccine composed of a mixture of bacterial products and brain proteins. In the meantime, those with the disease will receive more effective treatment as we better understand the immune system and the nervous system function, and develop more sophisticated imaging. This will lead to patient-specific treatment for MS. People who do come down with the disease will have it effectively cured. Perhaps we should say conquered rather than cured.

CHAPTER THREE

Alzheimer's Disease

I will never forget one particular scene with Dennis Selkoe, with whom I have worked for over thirty years. He was sitting with a seventy-eight-year-old woman with suspected Alzheimer's disease, who was accompanied by her daughter.

"When did you first suspect something was wrong?" he asked the daughter.

"I noticed it at our deli. My mother used to work the cash register, and she began mixing up the cash in the register, and she had trouble ringing things up. She would forget the price of a bagel and cream cheese. Then I found sales slips from the credit card machine in the trash."

"Anything else?"

"She couldn't find where the bagels were. She'd go to the refrigerator and couldn't find the brownies."

"Did she ever get lost?"

"She once got lost when she was driving to visit my aunt. We didn't make much of it because it was nighttime and snowing. But then she got lost on the way to the doctor. She called me, all upset, crying. I found her and took her to the doctor. That's when they first diagnosed Alzheimer's. The doctor gave her some medication and told her she couldn't drive anymore."

"I'd like you to remember three things," Selkoe said: "72 Main Street, a Red Ford, and a lampshade. Do you think you could do that?" he asked.

"I'll try," the woman said.

Dennis then performed a neurological exam, something I had done in countless MS patients. He checked eye movements, strength, coordination, and reflexes. As I watched, I didn't see any abnormalities on the exam.

"Your exam is normal," Dennis said. "Now let's see if you can you remember the three things I told you."

"Did you ask me to remember something?"

"Yes, I did. One of them was a car. Do you remember the other two?'

"Not really," the woman said.

"She always complains of not being able to remember things," her daughter said. "I tell her that there are only three things to remember: 'Stay healthy. Be happy. And look pretty.' She can never remember the first two, but she always remembers 'look pretty.'"

The woman laughed. The woman was my mother. I had brought her and my sister Rhoda to Boston for my mother to be examined by Dennis Selkoe.

My mother's symptoms began insidiously, first recognized by my sister at our family's deli in Denver. Once she started getting lost, the diagnosis was made. She manifested classic features of the first subtle and then alarming cognitive failure characteristic of Alzheimer's disease. She couldn't remember how to open up the safe in her house, although she had done it every day. She would turn on the water and forget to turn it off. She would ask the same question, over and over. At the deli she would sit at a table drinking coffee or having a bowl of soup, and when people came up and said hello to her, she would turn to my sister and ask, "Do I know them? Who are they? Do I like them?" She loved Hershey bars, and we discovered half a dozen unfinished Hershey bars in her purse. Early in her disease, we put her on a plane from Denver to Boston. At the baggage claim, I realized she couldn't remember which suitcase was hers. But she knew that something was wrong with her and said to me, "Howie, you're a famous brain doctor, can't you do anything for me?" Sadly, I couldn't.

We then hired a woman named Marie to be with her. Marie brought her to my son's wedding in New York, where my mother danced and said hello to people. Many didn't realize she was ill. She put on a good show, but the disease progressed nonetheless. My mother had always dressed perfectly and regularly went to the beauty shop. Now, she wouldn't change clothes unless someone told her the clothes were dirty, and she no longer wanted her hair done. When I visited her in Denver, it was hard to hold a conversation, so we would watch TV together. Once, when I took her out for dinner, I decided to explain my research to her as if she could understand it. It made me feel good. She smiled and enjoyed being with her son.

As the disease progressed further, she had trouble remembering me. When my sister said Howie was coming to visit, she asked, "Who's that?" She never became belligerent or angry. If anything, she became more childlike. She was affectionate. But what I remember most at that time was the blank look on her face—as if a light inside her brain had been turned off, and her brain was now empty. My sister and I were at her bedside when she passed away at the age of eighty-four.

I had begun investigating the immune system in Alzheimer's disease when my mother was diagnosed with the disease, and I have spent over a decade trying to develop a nasal vaccine for treatment. Although Alzheimer's disease was not originally considered an immune-mediated disease like MS, immunotherapy has become a major focus for the treatment of Alzheimer's disease.

THE DISEASE

As the population ages, more and more people will suffer from Alzheimer's disease. MS is a disease of young adults, and although Alzheimer's disease can occur in younger individuals, it is basically a disease of aging. Unlike MS, which begins as a relapsing/remitting disease and then can become progressive, Alzheimer's is progressive from the onset. Once it starts, it doesn't stop. Typically, Alzheimer's begins after age sixty-five with mild symptoms that last for two to six years. Moderate Alzheimer's with more obvious memory and cognitive problems follows and generally lasts another three to five years. This phase is marked by trouble with basic tasks of living, such as dressing and grooming. Patients with severe Alzheimer's require caregivers for all their activities; finally, the disease advances to the point where the patient becomes mute and is unable to walk, swallow, or control bladder or bowel function.

In the United States, Alzheimer's disease is the sixth-leading cause of death overall and the third-leading cause of death in seniors. More than five million people in the United States have Alzheimer's disease, and every five years, the risk of developing Alzheimer's disease doubles. All races and ethnic groups get Alzheimer's disease. All primates develop evidence of Alzheimer's as they age, as do many animals, including dogs, cats, and even polar bears.

Few diagnoses match the heartbreak of Alzheimer's. Beyond the anguish, Alzheimer's puts tremendous functional and financial stress on families. In the United States, sixteen million Americans provide unpaid care for people with Alzheimer's and other dementias, and many say the burden of caregiving damages their own health. By age eighty, three-quarters of people with Alzheimer's and other dementias will be admitted to a nursing home, compared to only 4 percent of the general population. One study found the annual average out-of-pocket spending for family members with Alzheimer's and other dementias was $61,522, compared to $34,068 for those families without dementia.

The number of deaths from Alzheimer's has almost doubled since 2000. Because of the aging population in the United States, the number of Americans with Alzheimer's is expected to almost triple by the middle of the century, from 5.5 million to 13.9 million. Not surprisingly, the cost of caring for Alzheimer's is also expected to skyrocket. By 2050, Medicare and Medicaid spending on Alzheimer's could top $1 trillion, four times what it is today.

When I was in medical school, Alzheimer's disease was not in the public's consciousness and was not recognized as the major health problem it is today. Most people are afraid of getting Alzheimer's disease, and as we age, everyone has occasional memory lapses that cause concern. I asked Dennis Selkoe what people should worry about as a first sign of Alzheimer's disease. He told me many people have trouble remembering names or dates, but this is usually in the realm of normal. Difficulty with episodic memory is more worrisome. Episodic memory refers to the memory of an event or an episode in daily life. It allows us to mentally travel back in time to an occurrence from the recent or distant past—details of a trip you took or remembering the last time you ate dinner at a restaurant, where you ate, who was with you, and what you ordered. It's especially worrisome if you can't remember an episode even when given clues about it. Another sign is asking a loved one the same question over and over. Getting lost is an indication of Alzheimer's, but this often occurs somewhat later in the disease.

Alzheimer's disease gets its name from a German psychiatrist named Alois Alzheimer, who cared for a fifty-three-year-old woman named Auguste D. Here is a description of Auguste D. by Alzheimer in the seminal talk he gave on November 4, 1906: "Her memory is greatly disturbed. If one shows her objects, she is able to name these correctly, but immediately afterwards has forgotten everything. While speaking, she often uses embarrassing phrases, single paraphasic expressions (milk-pourer instead of cup), sometimes becoming stuck and stopping. She no longer seems to know the use of certain objects. Her gait is undisturbed. She uses her hands equally well." He might have been describing my mother, who had the disease a century later. After Auguste D. died, Alzheimer examined her brain and reported a type of pathology that did not fit into any known category. He described what are now the classic features of the disease that bears his name: plaques and tangles. These form the basis of the crime scene in Alzheimer's disease.

THE CRIME SCENE

What is striking when you hold the brain of someone who has died of Alzheimer's is how shrunken the brain is—so extensive that it can be seen with the naked eye. Under the microscope, the cause for the brain shrinkage becomes obvious. There is enormous damage to the neurons, disrupting networks that are crucial for memory and thinking. The damage to neurons is associated with two of the classic features of Alzheimer's disease: plaques and tangles. The plaques are outside the nerve cells; the tangles are within them. There are millions of plaques in the brain tissue of a patient with Alzheimer's.

The plaques at the crime scene are made up of a protein called beta-amyloid, or amyloid-beta. A-beta is a protein containing anywhere from thirty-eight to forty-three amino acids. All proteins are made up of amino acids, which are like letters of the alphabet; whereas, letters of the alphabet combine to form different words, amino acids combine to form different proteins. A-beta is produced throughout the brain and normally dissolves when placed into solution, like sugar dissolving in a cup of hot tea. But with age and other factors, A-beta forms clumps that resist being broken down, and these clumps are the primary component of the amyloid plaques.

Alzheimer's is not the only disease in which A-beta is implicated. Sometimes, the A-beta aggregates are found in blood vessel walls and cause a condition known as cerebral amyloid angiopathy, or CAA. Because of amyloid in blood vessels, there can be complications in trying to remove A-beta from the brain by using antibodies that target the A-beta. CAA is a cause of cerebral hemorrhages, which led to the fatal bleeding in the brain of former Israeli prime minister Ariel Sharon.

The second major finding at the crime scene, visible only under the microscope, is neurofibrillary tangles. These tangles are made up of a protein called tau, which is about four hundred amino acids in length. Unlike A-beta, which is outside the nerve cell, the tau protein is found inside the nerve cell body. Tau is made in all neurons, and its normal function is to

ALZHEIMER'S DISEASE

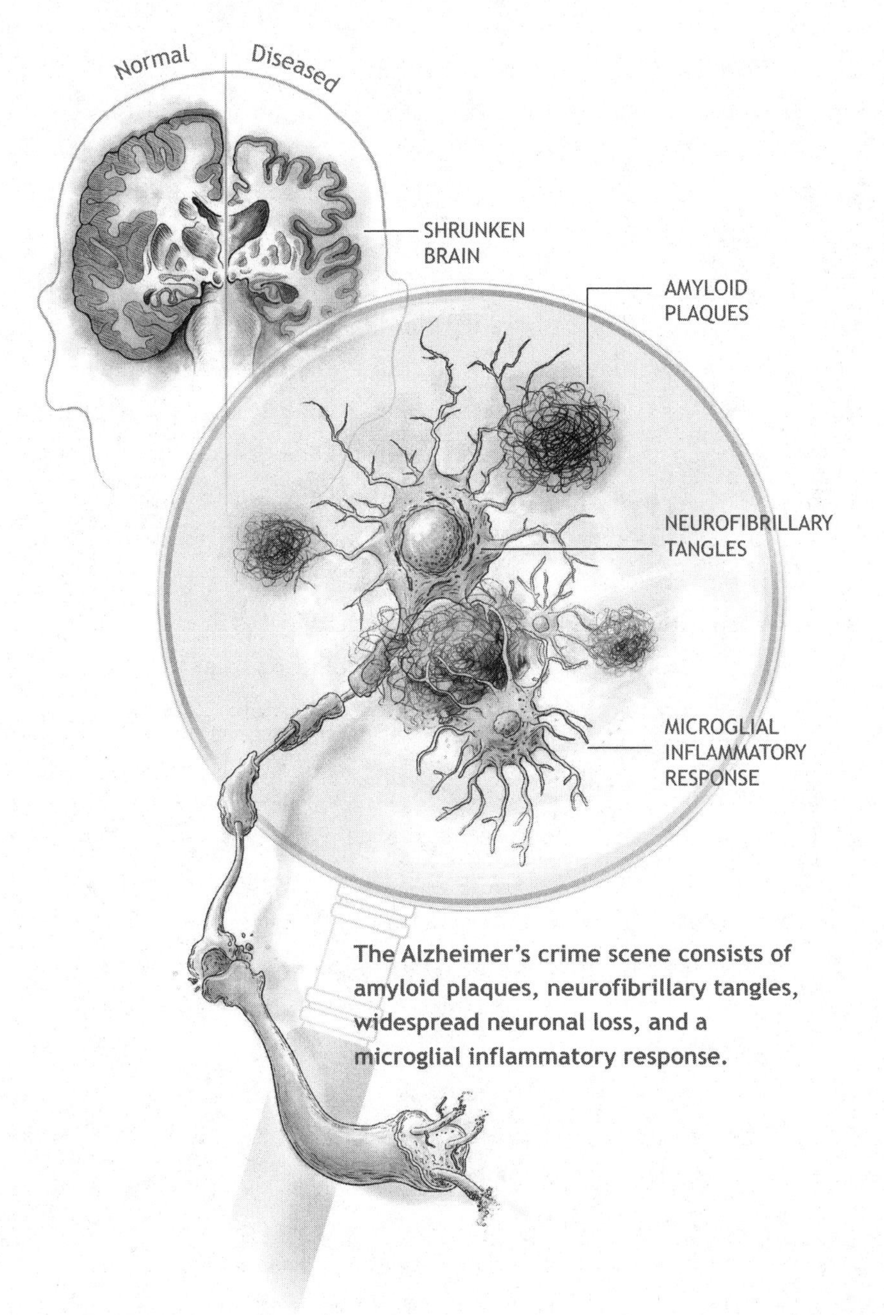

The Alzheimer's crime scene consists of amyloid plaques, neurofibrillary tangles, widespread neuronal loss, and a microglial inflammatory response.

stabilize the structure, or scaffolding, of the cell. In Alzheimer's disease, a portion of the tau protein aggregates to form structures called tangles. These tangles have a recognizable construction when viewed under the electron microscope, called paired helical filaments. The accumulation of tau inside the neuron causes damage to the neuron and is strongly linked to the clinical progression of Alzheimer's disease.

The presence of amyloid plaques and tau tangles activates local immune cells in the brain, which are a third major component of the crime scene. Immune cells in the brain may have a dual role in Alzheimer's disease: They can be beneficial and attempt to clear out abnormal accumulation of proteins like A-beta, or they can become toxic and increase the damage by producing inflammatory chemicals. The immune cells comprise microglia and astrocytes, and they play an important role not only in Alzheimer's disease but also in the other diseases we study at the Ann Romney Center—MS, ALS, Parkinson's disease, and brain tumors. Thus, the third major feature of the crime scene in Alzheimer's disease has to do with the brain's immune cells causing inflammation in the brain. In fact, inflammation and the involvement of microglia were noted in Alois Alzheimer's initial publication in 1907.

We know definitively that microglia are important in Alzheimer's because human genetics points to a key role for microglial cells in the disease. Certain gene mutations are associated with an increased risk for Alzheimer's disease, such as mutations on a gene called *TREM2*, which encodes for a receptor on immune cells, including microglia. A similar association occurs with the *ApoE4* gene, a major risk factor for Alzheimer's disease, which we will discuss below. As we will see, microglia are important not only in Alzheimer's disease and MS; the mutations in *TREM2* may also increase the risk for Parkinson's disease and ALS, as well as traumatic brain injury, which itself could be a risk for Alzheimer's disease.

A fragment of the TREM2 protein has been found to be increased in the spinal fluid of patients with Alzheimer's disease. This increase provides an opportunity to determine when microglial changes in Alzheimer's disease occur relative to the deposit of amyloid protein in the brain and the associated neuronal injury. Investigators studied a rare form of Alzheimer's disease in which a single gene can cause the disease; they found that TREM2 appeared after amyloid deposits but years before the onset of cognitive impairment and dementia. If there's a clear association between certain microglial genes and Alzheimer's disease, the question becomes how microglia are related to susceptibility to Alzheimer's disease. As with so many biological processes, especially in immunology, a particular cell type can adopt either beneficial or detrimental features, and this is particularly true with microglia.

Oleg Butovsky, who works in our center, has devoted his career to studying microglia, and he likens the different types to Dr. Jekyll and Mr. Hyde. Normally, microglia (Dr. Jekyll) function to maintain brain health and to deal with injury or abnormal proteins. They are called homeostatic, and they play a major role in shaping nerve cells and organizing the brain. However, when reacting to damage, some microglia can change into a toxic phenotype (Mr. Hyde) and cause harm themselves.

Studies show that in Alzheimer's disease, microglia can surround the plaques, and there they have a Mr. Hyde character. As we saw in our Progressive MS Initiative, microglial activation can be measured by PET imaging. Is the microglial activation seen by PET imaging in Alzheimer's disease good or bad? The answer is that it depends on the stage of the disease. In a longitudinal PET imaging study of people with Alzheimer's disease, activated microglia early in the disease course were associated with a better prognosis, whereas later in the disease course they were associated with a worse prognosis. It may very well be that when amyloid protein is deposited early, microglia attempt to clear the amyloid to protect the brain. They may then become overwhelmed and turn into toxic microglia that drive progression of the disease. These apparently toxic microglia are common to other diseases in the brain as well.

Thus, people may get Alzheimer's as they age not only due to an increase in the brain's deposition of A-beta but also because microglia become dysfunctional and lose their normal properties with age. When this happens, the microglia may be less able to clear A-beta from the brain. Indeed, in experiments in our laboratory, we found that microglia change as animals age, and if one injects A-beta into the brain of an older animal, the microglia are not able to clear the A-beta as well. Thus, if Alzheimer's disease were treated by either stimulating or dampening microglia, success would likely depend on the stage when the treatment was carried out. In early stages, it would be important to stimulate the ability of microglia to clear A-beta, whereas later in the disease, it would be important to decrease their toxic function.

Finally, and perhaps most important for cognition, neurons throughout the brain are affected, especially in specific regions that are crucial for memory, like the hippocampus. This is the reason my mother couldn't remember me. The loss of neurons leads to local shrinkage of the brain. Because of this, when you hold the brain of a patient with Alzheimer's, the ventricles, or CSF-filled cavities of the brain, appear enlarged, with gaps where the neurons used to be.

ORIGINS

Some of the most important clues about the origins of Alzheimer's come from genetic studies and one particular disease, Down syndrome. The unique facial features of people with Down syndrome have been recognized for centuries, and the first clinical description was made in 1866 by John Langdon Down, after whom the condition is named. But it wasn't until 1959 that the cause of Down syndrome was identified: an extra chromosome. Chromosomes contain all the genetic material in our bodies. Humans have twenty-three pairs of chromosomes (one from the mother and one from the father) for a total of forty-six chromosomes. In Down syndrome, instead of two of chromosome 21, there are three (Down syndrome is also known as trisomy 21). The extra chromosome 21 means there are extra copies of all the genes on chromosome 21.

Down syndrome is similar to Alzheimer's in that virtually all people with Down syndrome develop a crime scene in the brain that is indistinguishable from someone with Alzheimer's: enormous numbers of plaques and tangles. If you look under the microscope at the brain of a person with Down syndrome, it looks the same as that of someone with Alzheimer's disease. The clue to the connection was discovered in 1985, when the material in the plaques of both Alzheimer's and Down syndrome patients was found to be composed of beta-amyloid. Beta-amyloid comes, in turn, from APP, the amyloid precursor protein, and the gene that encodes APP is on chromosome 21. Thus, people with Down syndrome make more A-beta because they have an extra gene for APP, its precursor. The earliest deposits of A-beta in the brains of people with Down syndrome can be seen by age ten to twelve, but their blood levels of the A-beta protein are elevated at birth. At the Down crime scene, when the A-beta deposits first appear, they have not had time to clump, and there are also no tau tangles or microglial activation. Those reactions come later. Down syndrome has given a major clue to the origins of Alzheimer's and has led investigators to focus on the amyloid precursor protein and its natural product, A-beta.

One of the most basic biologic processes is cutting proteins into pieces, which is like making a smaller word from a big word. Enzymes, the body's special biologic scissors, are critical for normal biologic function because they accelerate chemical reactions. There are three known scissors that cut the amyloid precursor protein (APP) into pieces, and each cuts APP at a unique place. These scissors, or enzymes, are called alpha-, beta-, and gamma-secretase. One could have named them alpha-scissors, beta-scissors, and gamma-scissors. When APP is cut, one of the fragments created is A-beta, which, as we learned, can clump outside the cell, become toxic, and initiate the process of Alzheimer's disease. As mentioned above, A-beta can have anywhere from thirty-seven to forty-three amino acids, depending on exactly how APP is cut.

The most toxic form of A-beta is called A-beta-42. Like the other A-beta proteins, it is created by two scissor cuts of the APP protein—first by beta-secretase and then by gamma-secretase. This results in A-beta-42 because it has forty-two amino acids. Gamma secretase can cut at a different site as well, resulting in A-beta-40, which has forty amino acids, is more abundant than A-beta-42, and is less toxic. About 90 percent of the A-beta we normally produce is the less toxic A-beta-40.

Because the toxicity-prone A-beta-42 is created by the activity of beta-secretase and gamma-secretase, companies are trying to develop drugs that inhibit their activity and thereby decrease the levels of toxic A-beta-42 in the brain. These drugs are called beta- and gamma-secretase inhibitors, or BACE inhibitors. They work in animal models but have yet to be translated successfully to humans, in part because of side effects. On the other hand, cutting APP protein by alpha-secretase creates a smaller, nontoxic fragment and makes it incapable of generating the toxic A-beta-42. Thus, another theoretical treatment strategy is to increase the activity of alpha-secretase and thus prevent the generation of toxic A-beta-42.

To understand the origins of any disease, we always turn to genetics. If a gene is tightly associated with a disease, it must in some way be related to the cause. In some instances, the presence of a single gene explains the cause of the disease and is all that is needed for someone to have the disease. Huntington's disease is one such example. Woody Guthrie, the American folk musician, suffered from Huntington's disease. Like Alzheimer's disease, Huntington's disease is characterized by the death of nerve cells and the development of dementia.

Inheritance of Huntington's disease is related to the presence of a single gene and is called autosomal dominant. Thus, the children of someone with Huntington's have a 50 percent chance of contracting the disease, and 100 percent of children who inherit the abnormal gene from the parent who has the disease will contract Huntington's. Children who inherit the normal gene from the parent without the disease will not contract Huntington's. Like Alzheimer's disease, the deposit of an abnormal protein in the brain, called Huntington, damages cells.

Unlike Huntington's disease, most cases of Alzheimer's disease are not hereditary. However, a small subset of people who have early-age onset of AD have a strong genetic predilection and an autosomal dominant inheritance, like Huntington's. The major gene for the autosomal dominant form of Alzheimer's is called presenilin, so named because it causes a pre-senile form of the disease. Mutations in the gene are the most common cause of autosomal-dominantly inherited Alzheimer's disease, and account for approximately 1 percent of all Alzheimer's cases. The link of the presenilin gene to Alzheimer's was discovered by Peter St. George-Hyslop in Toronto in 1995.

In the movie *Still Alice*, Julianne Moore received an Academy Award for her portrayal of a middle-aged university professor who carried the abnormal presenilin gene and developed early Alzheimer's disease. In the film, some of her children elected to be tested for the gene, others did not. The big question, then, of course is if an abnormal presenilin gene has such a powerful effect that it causes Alzheimer's disease in all those that carry it, what is its function? The surprising answer was found by Mike Wolfe and Dennis Selkoe in 1999: presenilin is actually the gamma-secretase enzyme, which, as described above, is one of the two scissors responsible for cutting APP and creating the toxic A-beta-42 molecule, which we believe initiates the crime scene.

The brains of people with an abnormal presenilin gene who die for other reasons have telltale A-beta deposits, showing a direct connection between the gene and A-beta in the brain. Although it accounts for only a small fraction of Alzheimer's disease, presenilin can cause the disease in people in their thirties, forties, and fifties. The youngest case I have heard of is an eighteen-year-old woman who came down with Alzheimer's disease in high school and died at thirty-three. The largest known family with dominantly inherited Alzheimer's disease lives in the outskirts of Medellín, Colombia, where a family tree shows Alzheimer's going back more than two hundred years, with symptoms developing between ages forty and fifty. Disease in this family is caused by a presenilin gene mutation that increases the production of the toxic A-beta-42. Trials are underway to prevent the onset of clinical symptoms in family members by giving them an antibody that helps clear toxic beta-amyloid from the brain.

Although the presenilin mutation is the most common genetic factor that unequivocally causes familial Alzheimer's, other rare mutations have been identified in the *APP* gene. One such mutation identified in a Swedish family resulted in a lifelong increase of A-beta by as much as fivefold. Genetics thus suggest that the first criminal at the crime scene is the A-beta protein, which then forms clumps, leading to secondary tau alteration and microglial inflammation.

The amyloid buildup in the brains of people with Alzheimer's is the basis for the amyloid hypothesis of the disease, initially formulated in considerable part by my colleague, Dennis Selkoe. The amyloid hypothesis states that accumulation of the A-beta protein in the brain, specifically the accumulation of the A-beta-42 form, is the fundamental cause of Alzheimer's disease, and the driving force behind its initiation. There is overwhelming logic for the amyloid hypothesis, given all we know about the genetics and pathology of Alzheimer's disease, although the amyloid hypothesis has come under attack after numerous approaches to remove amyloid from the brain have failed in clinical trials.

One approach to decreasing toxic A-beta accumulation in the brain is to shut down the production of the toxic form of amyloid. As described above, secretase scissors cut the amyloid precursor protein in a way that results in both normal and toxic A-beta fragments. Thus, investigators have attempted to inhibit toxic A-beta production by inhibiting the activity of the gamma-secretase scissors. Several such inhibitors have been tried in Alzheimer's and have either failed to slow the cognitive decline or have had unacceptable toxicity. Studies with other inhibitors, however, are ongoing.

ApoE4

In addition to the presenilin gene that directly causes Alzheimer's disease in families, there are genes that increase the risk for the development of

Alzheimer's disease in the entire population. The most important one is called *ApoE4*.

The *ApoE* gene provides instructions for making a protein called apolipoprotein E, which combines with fats in the body and participates in packaging cholesterol and other fats to carry them through the bloodstream. *ApoE* has three major forms known as *ApoE2*, *ApoE3*, and *ApoE4*. The most common is *ApoE3*, found in about 70 percent of the population. Some 20 percent of us have an *ApoE4* gene, and only about 5 percent of us have an *ApoE2* gene. *ApoE4* is one of the major factors being investigated in our attempts to understand and cure Alzheimer's disease; it is a major component of the crime scene.

ApoE was discovered in 1993 when Warren Strittmatter, a postdoctoral fellow in Allen Roses's laboratory at Duke University, decided to place synthetic A-beta on a filter and pour human spinal fluid over the filter to see what proteins stuck. He wanted to know whether there was anything in the spinal fluid that bound to A-beta and thus might be involved in Alzheimer's disease. Strittmatter discovered that one of the proteins that bound to A-beta on the filter was ApoE. He told this to his mentor Alan Roses, a geneticist, who by chance had been working on chromosome 19, the very chromosome where the *ApoE* gene is found. It was then very easy to genetically test Alzheimer's patients to see whether the *ApoE* gene was linked to their disease. In a major discovery for the field, Roses and Strittmatter found that Alzheimer's patients were much more likely to carry an *ApoE* gene than people without Alzheimer's. In other words, carrying *ApoE4* increased your likelihood of getting the disease. Of the three *ApoE* gene forms, *ApoE4* is a risk factor present in perhaps 40 to 50 percent of all people who develop Alzheimer's, whereas *ApoE2* is protective against the disease. *ApoE3* is considered neutral.

To discover how the *ApoE4* gene increases the risk of Alzheimer's in the general population, David Holtzman at Washington University in St. Louis conducted experiments with mice that didn't have an *ApoE* gene of their own. Through the wonders of genetic engineering, he inserted the genes for either human *ApoE4* or human *ApoE2* into them. He found that the *ApoE4*-expressing mice had a defect in the clearance of A-beta from the brain. As a result, the A-beta aggregated more. Mice with *ApoE2* had less A-beta accumulation and aggregation; thus, *ApoE2* is protective. A single copy of the *ApoE4* gene increases the risk of Alzheimer's by three to four times. Two *ApoE4* genes, one from your mother and one from your father, increases your risk of Alzheimer's as much as fifteenfold.

Looking at common diseases such as diabetes, hypertension, and atherosclerosis, Holtzman says he doesn't know a single gene that affects us humans as much as *ApoE*. As many as 50 percent of Alzheimer victims worldwide appear to have the disease because of *ApoE4*. Remember, unlike a mutant presenilin gene, *ApoE4* does not cause Alzheimer's disease in everyone who has it. It is a risk factor, the way smoking is for lung cancer. You don't have to smoke to get lung cancer, but it's much more likely if you do. It appears that *ApoE* may be linked to Alzheimer's through its effect on microglial function. In clinical trials, patients may be stratified based on *ApoE4*.

There are programs underway to convert the *ApoE4* protein in the brain to a form that does not increase the risk of Alzheimer's. The *ApoE* protein has 299 amino acids, and the difference between the major types of *ApoE* is a single amino acid, like the difference of a single letter in this paragraph. As discussed previously, amino acids are like letters of the alphabet and combine in various ways to form particular proteins. *ApoE4* has the amino acid arginine at position 158, while the protective *ApoE2* has the amino acid cysteine in that spot. Thus, the placement of one amino acid can change the properties of a protein. Changing one letter can change the entire meaning of a word, like in the words *love* and *live*. The same can happen when a single amino acid is changed in a protein.

AN A-BETA VACCINE

A major milestone in the study of Alzheimer's disease occurred in 1995. Although imperfect, animal models of human disease are extraordinarily helpful in understanding disease mechanisms and devising treatments. Virtually all diseases have some animal counterparts. The most valuable counterparts are mice, as we can easily manipulate their genes. The first mouse model for Alzheimer's disease was created in 1995 by inserting a gene from a familial form of AD into a mouse. This transgenic mouse produced human A-beta throughout its life, which led to an increase in the toxic A-beta-42 in the brain. The animals developed amyloid plaques, had loss of neurons, and an increase in microglial cells and scarring. Interestingly, there were no tau tangles, the second major feature of the Alzheimer's crime scene, which are generally not present in rodent tissue. Disease in the brain wasn't observed until the mice were about nine months old (we consider two to three months an adult mouse, an age we use for most mouse experiments), and then amyloid deposits accumulated progressively with time, just as in people with Alzheimer's disease.

Development of Alzheimer's transgenic mice set the stage for an experiment that was published in *Nature* four years later in 1999. It revolutionized the field by creating an approach for treating Alzheimer's disease that was based on the immune system. Dale Schenk and his colleagues at Elan Pharmaceuticals in San Francisco postulated that by immunizing Alzheimer's transgenic mice with A-beta protein, they would create an immune response against A-beta that could clear the A-beta in the brain. Thus, they mixed the A-beta-42 protein in an adjuvant that boosts the immune system, called Freund's adjuvant. Freund's is the same adjuvant we use to induce the MS-like EAE disease in mice when we inject a myelin protein.

Schenk and his colleagues found that immunization with A-beta-42 in adjuvant given prior to A-beta plaque development, when the mice were only six weeks old, markedly reduced the plaque levels later in life, at thirteen months. They then treated mice that were twelve months of age when they already had plaque deposits. Again, they found a positive effect. Fewer plaques were observed in these mice when they were analyzed at fifteen and eighteen months of age. The scientists had shown prevention of A-beta plaque formation, less dysfunction of neuronal processes, less scarring, and less microglial inflammation. Importantly, there were no signs of side effects or other damage in the animals.

Researchers have developed tests for cognitive decline in mice that mimic those done with Alzheimer's patients. To test working memory, researchers use mazes. Schenk's studies showed that not only were the plaques cleared but mice treated with this approach did better in water maze tests. Although the brains of mice are far less complex than those of humans, these well-established tests showed that clearing the A-beta actually improved mental function. For executive function, mice need to understand cues such as odor or texture to know which bowl contains buried food or to adapt when rewarded and unrewarded behaviors are switched. Attention span tests require a mouse to respond in different ways to different flashing lights to receive a food pellet.

The success of A-beta-42 immunization was assumed to be related to an immune response against A-beta, although the mechanism was never reported in the first paper. Important controls, however, were done. Immunization with adjuvant alone or with a different peptide had no effect. Thus, the effect was related to a specific immune response against A-beta-42. There were two possibilities. The first was the induction of anti-A-beta-42 antibodies, postulated by Schenk. The second was the induction of T cells specific for A-beta-42, like the T cells specific for myelin that we induce in the MS EAE model. The formal proof of these possibilities would be to show that injecting antibodies against A-beta-42 or anti-A-beta-42 T cells in PDAPP Alzheimer's mice would have the same effect. Evidence for the role of antibodies occurred one year later, in 2000, when the Elan investigators reported that injecting anti-A-beta-42 antibodies into the bloodstream of the PDAPP mice entered the nervous system, bound to A-beta plaques, and reduced A-beta levels in the brain.

These exciting results, of course, immediately raised the possibility that a similar approach in humans could work in Alzheimer's disease and become the first disease-modifying treatment. Elan Pharmaceuticals moved quickly to perform human trials. The concept of vaccination is nothing new and is something that immunologists have applied successfully to infectious disease. Vaccination with a noninfectious form of a virus such as poliovirus or, in the case of smallpox, the cowpox virus, leads to developing protective immunity. In the case of Alzheimer's disease, the first vaccinations were done by Elan Pharmaceuticals with A-beta protein, the accumulation of which was assumed to be the culprit in Alzheimer's. Triggering an immune response against A-beta would clear amyloid from the brain. No side effects were observed in these mouse studies, and two Phase 1 studies in approximately eighty people with Alzheimer's disease turned up no drug-related concerns or side effects. Unlike in mice, it wasn't possible to use Freund's adjuvant to boost the immune response in people, so another adjuvant, called saponin QS-21, was used. Based on these results, a Phase 2 trial, called AN-1792, was initiated in 372 patients. Unexpectedly, it had a complication that had not been seen in animal studies or in the Phase 1 human trials. Six patients developed encephalitis, or brain inflammation, and the study was stopped. No one knows for sure why this complication was observed in the Phase 2 trial and not in Phase 1. Most believe it was related to the type of immune response that was induced by the adjuvant. Immune studies suggested that in the Phase 1 trial, a Th2 immune response was induced, whereas in the Phase 2 trial, a Th1 immune response was induced. We know from our studies in MS and the EAE animal model that the type of T cell immune response induced is crucial for causing encephalitis. A Th1 immune response would induce encephalitis, whereas a Th2 immune response would not. Different adjuvants can drive a Th1 versus a Th2 immune response. However, the same adjuvant, QS-21, was used in both the Phase 1 and Phase 2 trials. The only apparent difference was that a detergent called polysorbate-80 was used during manufacturing to stabilize the vaccine in the Phase 2 trial, and it is possible that this could have shifted the type of T cell immune response. It is amazing that such a small variable could have had such a profound effect.

At the time the paper on vaccination with A-beta-42 protein was published in *Nature*, Dennis Selkoe and I were working on another vaccination approach for Alzheimer's disease. Our approach was to give the A-beta-42 protein intranasally. The idea stemmed from my interest in the mucosal immune system as a route through which immune therapy could be given orally or nasally. I was particularly interested in the oral and nasal routes because they are noninvasive natural routes to the immune system, as opposed to an injection with a needle, which is artificial. I postulated that nasal A-beta-42 protein would decrease the inflammatory reaction in the brain of patients with Alzheimer's.

Like Schenk, we used Alzheimer's mice that developed A-beta plaques. I remember sitting with Dennis and looking under the microscope at the brains of mice that were given nasal A-beta versus those given a control protein. After looking at the slides and breaking the code, Dennis was jubilant. He gave me a hug and said, "Howard, we have an effect here." We indeed had an effect. We had reduced the amount of A-beta in the brain by giving the A-beta-42 nasally. I knew the Schenk paper was under review at *Nature*, soon to be published, and I contacted the editors at *Nature* and told them of our similar findings. I asked whether our results could be published alongside the Schenk article. However, it was too late, and we ultimately published in 2001 in the *Annals of Neurology*.

We were confronted with the same question as Schenk. How was our nasally administered A-beta-42 working? We found decreased inflammation in the brain. Like Schenk, we also found we were inducing antibodies against A-beta-42, and there was an association between the development of these antibodies and the decreased A-beta in the brains of the mice. If we were inducing antibodies in a different way, we wondered if our approach was safer, given that we were using a natural mucosal route.

The nasal approach was patented by a company I had founded called Autoimmune, which was developing the mucosal route for the treatment of autoimmune and inflammatory diseases. Elan bought the rights to nasal A-beta but never developed the approach, as they were conducting injectable vaccine trials. I've always wondered if the timing had been different and nasal A-beta-42 had been tried in humans, whether it would have had positive effects with no toxicity and been the first immune modulating drug for Alzheimer's disease.

Dennis and I immediately moved forward to test nasal A-beta vaccine in Alzheimer's disease. Experimental drugs given to humans require an investigational new drug application (IND), so Dennis and I had A-beta-42 made that could be given to humans, and we filed an IND with the FDA. The FDA, by law, must give a response to new INDs within one month, and soon we were on the phone with the FDA discussing nasal A-beta-42 for treating Alzheimer's disease. As it turned out, while our application for nasal A-beta-42 was pending at the FDA, the encephalitis complication occurred with the injectable vaccine. Consequently, the FDA put all studies of vaccination with A-beta-42 on hold. We could go no further, and there was little we could do. We had observed no toxicity in our animal studies, but of course Elan had observed no toxicity in their animal studies of the injectable vaccine, either. So, we had A-beta-42 ready for human use, and all we could do was turn it over to scientists in the lab to use on mice.

PROTOLLIN

The occurrence of encephalitis in people immunized with A-beta-42 led me to a series of experiments that opened up a totally new approach for treating Alzheimer's disease using an adjuvant called Protollin. My hypothesis for what had caused the encephalitis with the injectable A-beta-42 was that immunization with the A-beta-42 created a situation similar to what happens in MS. In MS, T cells that recognize brain proteins leave the bloodstream, enter the brain, and cause MS attacks by damaging the myelin sheath. In the laboratory, we were studying T cells against A-beta and found that people had T cells against A-beta, and these T cells were increased in patients with Alzheimer's disease. I hypothesized that immunization of Alzheimer patients with A-beta-42 plus adjuvant stimulated A-beta T cells in the blood of Alzheimer's patients to enter the brain and cause encephalitis. I decided to perform animal studies to better understand what had happened.

The first question I asked was whether one could immunize Alzheimer's mice with A-beta-42 in adjuvant and induce encephalitis, just like what had occurred in Alzheimer's patients during the Phase 2 trial. The experiment failed: no disease. The second question I asked was whether the Alzheimer's mice were more susceptible to EAE, the mouse model of MS. This might explain why people with Alzheimer's were, for some reason, more susceptible to encephalitis after injection with A-beta protein. We decided to immunize Alzheimer's mice with a brain protein in an adjuvant and measure the degree of paralysis. Would the Alzheimer's mice get worse MS? The experiment was negative: there was no difference in the severity of paralysis between mice that had A-beta in the brain versus those that didn't. I had another failed experiment. However, when I looked at the brains of mice with Dan Frenkel, who was running the experiments, we made a startling discovery: A-beta plaques had been cleared from the brains of the Alzheimer mice that were given MS. We had unexpectantly found that we could clear A-beta from the brain with an immunization that didn't involve A-beta-42. We knew the observation was real because the findings were robust, and more importantly, they were repeatable.

Like all observations, the central question was, Why did this happen? Perhaps, for some reason, we were inducing antibodies against A-beta although we hadn't immunized with A-beta. If that was the case, it would be far less interesting. The field already had antibodies against A-beta to test. Although we couldn't find antibodies against A-beta in the mice, it wasn't 100 percent proof. Perhaps some antibodies were undetectable, or our assay was imperfect. So, we turned to genetically engineered mice that were incapable of making any antibodies because they had no antibody-producing cells. We mated these antibody-deficient mice to Alzheimer's mice. We now had animals with human A-beta in the brain who were genetically incapable of making anti-A-beta antibodies. To our excitement, when we immunized them to get MS, they still cleared A-beta from the brain. Thus, we had discovered a unique way to clear A-beta from the brain in a mouse model of Alzheimer's that didn't depend on making antibodies against A-beta. This was an exciting discovery, but, of course, you can't treat Alzheimer's patients by injecting them and causing an MS-like state.

The immunization given to Alzheimer's patients that led to encephalitis was given by injection. As discussed, I had always been interested in natural routes of stimulating the immune system via the nasal and gut mucosa, as these sites have a unique immune system. So, I asked the question of whether one could stimulate the immune system via the nasal route in a way that could clear the beta-amyloid from the brain without causing disease.

We found a company in Canada that made nasal vaccines and had a particular adjuvant that had been tested for flu vaccinations. It was called Protollin. To our joy we found that when we gave the Protollin nasally, immune cells called monocytes were activated and traveled to the brain, where they localized to the brain amyloid plaques and cleared the plaques. Not only were plaques cleared but when we performed memory testing, we also found that the animals had better memory function.

We then carried out an experiment in which we used Protollin to treat mice in a preventative mode. To effectively treat Alzheimer's, treatment would have to be given early and continue for a long time. Using Alzheimer's mice, we treated them with Protollin weekly, beginning when they were five months old, for a full eight months. Would prolonged treatment started early have an effect? This question was very important later, when it became clear that trials to remove A-beta in people using antibody alone worked best when started very early. Indeed, as we will show, there are current trials in which cognitively normal people are treated with an antibody to remove A-beta. If our nasal vaccine worked when given early to mice, it could be given to cognitively normal people. The experiment worked: we found a dramatic reduction of amyloid deposits in the Alzheimer's mice treated for eight months with Protollin. Eight months is a long time to treat a mouse, but it's the only way we could test the system in a way that is analogous to human disease.

When we examined the brains of the mice treated with intranasal Protollin, it became clear that cells activated by the adjuvant in the blood traveled to the brain, attached to the A-beta, and cleared it. We found the Protollin in the nasal cavity but not in the brain; thus, it appeared we were stimulating immune cells outside the brain, which then entered the brain and cleared the amyloid.

These results were very exciting, as we had identified an immune treatment for both the potential prevention and treatment of Alzheimer's disease that did not require antibodies against A-beta. This led us to hypothesize that immune cells called innate cells play an important role in Alzheimer's and that targeting them would be a unique approach for treatment. The innate immune cells in the brain are microglial cells, and those outside the brain are called monocytes. Both have scavenger function and can "eat" A-beta and clear it from the brain. We hypothesized that gradual accumulation of amyloid when people age is related to a defect in the ability of microglial cells in the brain and scavenger cells in the blood to clear amyloid from the brain and prevent brain damage. With Oleg Butovsky, we identified unique signatures for microglial cells and monocytes. Thus, we could distinguish the two cells and target them separately.

I spoke to Dennis about our results and told him I was going to apply for an NIH Director's Transformative Research Award. Such grants are awarded to "exceptionally innovative and unconventional research projects" that are inherently risky but have the potential to create or overturn fundamental paradigms. NIH transformative grants are extremely competitive, and only very few are awarded each year. Dennis encouraged me but told me not to get my hopes up. I asked him to write the section of the grant related to A-beta, and I sent it in. Luckily, I got the grant. The underlying hypothesis for this work is not idle speculation. We know that microglia and monocytes play a role in Alzheimer's disease, because susceptibility genes for Alzheimer's disease are expressed on the microglial surface. Moreover, genetically altered animals in which the movement of monocytes into the brain is impeded have more amyloid deposition. Furthermore, the clearance of amyloid with antibodies against A-beta antibodies appears to depend on the stimulation of microglia to clear A-beta. Thus, I think we have the right targets.

Francis Collins, the NIH director, held a competition for Transformative Research Award winners to create a video that described their scientific approach for the lay public. He participated in judging the entries. I took the "Cups Song" from the movie *Pitch Perfect* and filmed the lab performing the cups routine with lab beakers as I showed pictures of older people and explained our science. Dennis joined me doing the cups routine. It won first prize. You can watch it at:

| https://www.youtube.com/watch?v=iA0jVd_tDeo |

It took fifteen years from the time we published our first paper on Protollin until we were ready to test it in Alzheimer's disease. I remember sitting with Ann Romney during one of her MS checkups when she asked what was new in the lab. I told her about our MS research and then causally mentioned that we were also working on a nasal vaccine for Alzheimer's disease. Her mouth fell open. I then told her about work we were doing on ALS. At that moment the Ann Romney Center for Neurologic Diseases was born. We decided to "go big" and create a center focused on five of the most difficult neurological diseases. I spoke to Dennis Selkoe, and he immediately came on board. Ann had hoped to advocate for neurological diseases as First Lady of the United States, but it was not meant to be.

I had been trying for over a decade to test Protollin in Alzheimer's disease. Finally, the stars aligned. With philanthropic funds from the Ann Romney Center igniting the process, I found commercial partners to manufacture the Protollin—Jingwu Zhang, a scientist who'd been a postdoctoral fellow in my lab in the early 1990s and who went on to found I-Mab biopharmaceutical company, and Joe Zimmermann, of Biodextris in Canada. We have completed manufacturing, submitted an investigational new drug application (IND) to the FDA, and are now on track to test Protollin in Alzheimer's disease. We are in position to take an exciting and potentially important "shot on goal." I know how difficult it is to translate mouse findings to people. Nevertheless, to paraphrase Winston Churchill, "Never, never, never, give up!"

A-BETA ANTIBODIES TO TREAT ALZHEIMER'S

Although the A-beta immunization trial that caused encephalitis had to be stopped, a number of trial participants had been treated, and investigators analyzed the results of the aborted trial to determine if they could observe any positive effects. Christophe Hock and Roger Nitsch at the University of Zurich assessed cognitive function in thirty patients who had received the injected A-beta vaccine. They found that twenty of the patients made antibodies against beta-amyloid, and that patients who generated these antibodies had slower rates of decline in both cognitive function and activities of daily living, as indicated by the Mini–Mental State Examination, compared to patients without such antibodies. They also found these effects in two patients who had transient episodes of encephalitis. The authors concluded that antibodies against A-beta plaques can slow cognitive decline in patients with Alzheimer's disease, although in the aborted Phase 2 trial, one could not judge cognitive performance because only two to three of the planned five injections had been administered.

The aborted trial was also analyzed by the study investigators, but some years after those two to three injections had been given. A post hoc analysis—analyzing the results of a failed trial to see if any positive signs could be found and if they occurred in a subgroup of patients—concluded that the neuropsychological tests favored antibody responders. Post hoc analysis is classically done in failed trials to search for signals of efficacy, but they can be misleading, as the trials were not initially designed to test the variables analyzed in the post hoc analysis. This may be one reason why Phase 3 trials fail when they are based on post hoc analysis of Phase 2 trials. Nonetheless, the benefits of identifying responder and nonresponder groups post hoc to help with additional trials cannot be ignored. For instance, this post hoc analysis revealed that in patients who had spinal fluid examinations, there was a decrease of tau protein in the spinal fluid of patients who developed antibodies against A-beta as compared to placebo subjects. This suggested a biological effect, as tau may, in the end, be a key molecule mediating cognitive dysfunction in Alzheimer's. However, long-term follow-up of the eighty patients who were immunized with AN-1792 showed no benefit either in terms of survival or major delay in the time it took to develop severe dementia.

After extensive analysis, the aborted A-beta immunization trial provided the following pieces of information. First, postmortem examination of the brains of patients who were immunized against A-beta showed an effect on A-beta plaques and other parts of the crime scene, such as tau. Second, side effects were observed, which had to be taken into account in further trials. And third, although there was no evidence of a dramatic effect on cognition, some positive signals were seen. Was there a path forward to target A-beta in Alzheimer's disease? The answer was a resounding yes. Instead of immunization with A-beta to create antibodies, the stage was set for a series of trials conducted in thousands of patients in whom antibodies against A-beta were given by intravenous infusion.

Treating Alzheimer's disease by infusing antibodies against A-beta was based on another paper published by Dale Schenk and coworkers, this time in *Nature Medicine.* In that paper, he bypassed immunization as a method to induce antibodies to clear A-beta in animals, and instead directly gave the antibodies by infusion. Schenk's team found that antibodies against A-beta administered directly into the bloodstream were able to enter the brain, localize to the A-beta plaques, and induce clearance of the preexisting amyloid. He also carried out experiments in the test tube in which he found that antibodies against A-beta triggered microglial cells to clear plaques by engulfing the A-beta. Thus, although the antibody was administered as a therapeutic agent, it appeared that downstream microglial cells might have been involved in clearing the A-beta.

With some positive signs from the aborted active A-beta immunization trial and Schenk's paper in *Nature Medicine*, the stage was set for human trials in which antibodies against A-beta were given intravenously to people. This held the promise of treatment for Alzheimer's and would directly test Dennis Selkoe's amyloid hypothesis that amyloid was the driver of Alzheimer's disease. If the hypothesis was true, removal of amyloid should have a beneficial effect.

It took almost ten years from the aborted A-beta immunization trial for the trials of intravenous anti-A beta to be completed. On January 23, 2014, the results of four large Phase 3 trials of two different anti-amyloid antibodies were reported in the *New England Journal of Medicine*.

The first antibody to be tested was called bapineuzumab, which was tested in mild to moderate Alzheimer's disease in two independent trials. The trials were funded by Janssen and Pfizer. Both trials were double blind, randomized, and placebo controlled. Because *ApoE4* is a risk factor for Alzheimer's disease and people with *ApoE4* have a worse course of the disease, both *ApoE4* positive and negative patients were studied. One trial enrolled 1,121 carriers of *ApoE4*, and the other 1,331 non-carriers. Each study took five years to complete and involved a total of over five hundred clinical sites. This was all done at great cost and effort.

The subjects were given an intravenous infusion every thirteen weeks for eighteen months. The primary outcome measure was cognition. Secondary outcome measures included findings on brain imaging of amyloid and spinal fluid concentrations of tau protein. For every Phase 3 trial, the FDA requires that a primary endpoint or primary outcome measure be identified before the trial begins by which the trial is then deemed a success or failure. Sadly, the results in both the carrier and non-carrier studies were negative, meaning the primary outcome measure was not met. Some believed the trial failed because the dose of the antibody had been kept very low to prevent a side effect called amyloid-related imaging abnormalities (ARIA), or focal areas of swelling seen on MRI. Indeed, the appearance of ARIA at higher doses and in those who were *ApoE4* carriers led to the discontinuation of the highest dose being tested. There were, however, some differences in amyloid brain imaging and spinal fluid tau concentrations between treatment groups; thus a minor therapeutic signal was observed. The authors concluded that amyloid accumulation probably starts many years before symptoms occur and that beginning treatment after dementia develops may be too late to affect the clinical course of the disease.

In the same issue, the *New England Journal of Medicine* reported the results of two other Phase 3, double-blind trials of another antibody against beta-amyloid, called solanezumab, funded by Eli Lilly. Over two thousand patients received an intravenous infusion every four weeks for eighteen months. The trials were named EXPEDITION and EXPEDITION2. Again, the primary outcome measure was cognition. Unfortunately, as with bapineuzumab, the trials of solanezumab failed. Treatment did not improve cognition or functional ability, although there were positive trends observed that did not achieve statistical significance. Some postulated that it didn't work because it didn't bind to the form of A-beta that in animal models was shown to impair memory. With the failure of these four trials, an editorial in the *New England Journal* by Eric Karran and John Hardy asked, "Antiamyloid Therapy for Alzheimer's Disease—Are We on the Right Road?"

Despite the cost, Eli Lilly felt it was on the right road and persisted. They carried out a third trial called EXPEDITION3 in which they used the same antibody, solanezumab, that they used in EXPEDITION and EXPEDITION2. This time they focused on patients in an earlier stage of disease. Just as important, they confirmed that all patients treated had amyloid in their brain as confirmed by brain amyloid imaging. In previous antibody trials, as many as 25 percent of patients diagnosed with Alzheimer's enrolled in a trial did not have increased amyloid levels in the brain, as measured by imaging. An antibody against amyloid could not be expected to have an effect if the patient didn't have amyloid in the brain. Thus, the EXPEDITION3 trial, which was done in people with mild dementia, required patients to have an appreciable amyloid load as seen by imaging.

The results were published four years later on January 24, 2018, again in the *New England Journal of Medicine*. Everyone hoped the trial would be successful. Unfortunately, solanezumab at a dose of 400 mg given every four weeks did not significantly affect cognitive decline. Although there was a trend, it did not reach statistical significance. Perhaps there was too much amyloid in the brain or simply that treating mild Alzheimer's was too late. M. Paul Murphy wrote in an editorial that accompanied the results of EXPEDITION3, "We may very well be nearing the end of the amyloid-hypothesis rope, at which point, one or two more failures will cause us to loosen our grip and let go."

Needless to say, the failure of these major trials was a great disappointment to the Alzheimer's community. Some argued that therapy needed to be given to patients with amyloid deposits who were cognitively normal. Such a trial indeed is underway, headed by Reisa Sperling at our hospital in Boston. The trial is called the A4 study: <u>A</u>nti-<u>A</u>myloid Treatment in <u>A</u>symptomatic <u>A</u>lzheimer's study. Reisa is using Lilly's solanezumab antibody and quadrupling the dose because the results of the previous trials were negative. For the study, seven thousand clinically normal individuals between sixty-five and eighty-five were screened for the presence of elevated fibrillar brain amyloid levels. If these were found, patients were enrolled in a three-year placebo-controlled trial of solanezumab, given by infusion every four weeks. In addition to patients who have abnormal amyloid brain scans, an additional group of patients without elevated amyloid will be followed to determine their risk.

Another approach based on the theory that antibodies against A-beta would be beneficial in Alzheimer's disease involved treatment with natural antibodies called intravenous immunoglobulin, or IVIG. IVIG is a mixture of antibodies found in donated human blood, which are pooled and used to treat a number of conditions, including autoimmune diseases. It has even been tried in MS. IVIG was tested in Alzheimer's disease based on the idea that it may contain natural antibodies against A-beta and could easily be tested because it was safe, as it is in common use.

One of the theories related to Alzheimer's postulated that there are natural antibodies against A-beta in the blood, and that people with Alzheimer's may have a decrease in these antibodies. Investigators showed that there are antibodies against A-beta in human blood IVIG. Two small studies, one in five patients and another in eight, suggested there was a benefit of IVIG as measured by increased A-beta in the serum, suggesting it was being cleared from the brain.

Another approach to test this hypothesis was to determine whether patients who had received IVIG for other causes had any development of Alzheimer's disease—an epidemiologic approach to find factors related to Alzheimer's. Investigators searched a database of twenty million people over the age of sixty-five and found cases and controls. They reported that previous treatment with IVIG was associated with a reduced risk of developing Alzheimer's. Thus, the stage was set for larger IVIG trials.

An initial trial of fifty patients with mild cognitive impairment was carried out in which patients received five infusions of IVIG every two weeks. Cognitive and brain atrophy measures were performed. Positive results were seen on the Mini–Mental State Examination at one year but not at two years.

Based on the positive results in this initial trial, a definitive trial of IVIG in Alzheimer's disease was performed. A Phase 3, double-blind, placebo-controlled trial randomly assigned 390 patients with mild to moderate Alzheimer's to receive IVIG or low-dose albumin as a control. Unfortunately, no beneficial effects were observed in the two primary outcome measures: changes from baseline on the Alzheimer's Disease Assessment Scale–Cognitive Subscale (ADAS-Cog), which measures language, memory, and orientation and changes in the Alzheimer's daily living inventory, including eating, dressing, and basic hygiene. There were decreases, however, in plasma A-beta-42 among those receiving IVIG but not in controls. It is possible that the trial of IVIG, if started earlier, may have been of benefit, but this can only be established if further studies are performed.

As it turns out, investigators were not at the end of the amyloid hypothesis rope. The attention and hope of the Alzheimer's community next turned to an antibody called aducanumab, developed by Biogen. It had demonstrated the strongest results to date for antibody therapy against amyloid in a Phase 1 trial. Biogen reported dramatic results of the double-blind, placebo-controlled trial called PRIME. The results were published on September 1, 2016 in *Nature*, one of the most difficult journals in which to get a paper accepted, especially a clinical trial. The PRIME study was led by Al Sandrock of Biogen. The Phase 1 study was relatively small, involving a total of 165 patients. Forty received the placebo, while the other 125 received increasing doses of aducanumab. The study provided the first demonstration of a biological and clinical effect of an antibody against A-beta. Patients had prodromal or mild AD, and treatment with aducanumab reduced brain A-beta plaques, as measured by brain amyloid imaging, in a dose- and time-dependent fashion. The clearing of the A-beta was accompanied by a slowing of clinical decline as measured by clinical scores after one year. Although this was a relatively small Phase 1 study, Biogen elected to move directly to two Phase 3 studies of the drug in studies called ENGAGE and EMERGE. Each Phase 3 study involved over 1,300 patients in a double-blind, placebo-controlled trial.

I spoke with Al Sandrock, the chief medical officer at Biogen, who has been instrumental in developing drugs for neurological diseases, including ALS and MS. He told me that aducanumab had a better chance than the other monoclonal antibodies that failed in clinical trials because the monoclonal antibody was cloned from blood cells from humans who had been found to make antibodies against A-beta, rather than antibodies generated in mice.

Although other antibodies had failed, Sandrock, along with Cecil Pickett, his head of research and development, went to Biogen cofounder and Nobel Prize–winning geneticist Phil Sharp with the proposal. It was approved. It cost one million dollars to license. Al told me if the drug worked, it would go down in history as the best one-million-dollar investment ever made by a biotech company.

A boost for the possibility that an anti-amyloid antibody may ultimately work in Alzheimer's disease came with reports at the Alzheimer's Association International Conference in Chicago in July 2018. Another antibody against amyloid was being developed jointly by the Japanese company Eisai and Biogen. It was called BAN2401 and targeted amyloid-beta protofibrils. It showed positive effects in a trial of 856 patients with early Alzheimer's disease. The eighteen-month interim report showed that the highest dose was able to both reduce plaques and slow progression of cognitive decline. Some people were excited with the results; others were cautious because of the early stage of the drug's development. ABC news reported, "New Alzheimer drug shows big promise," while CBS news reported, "Alzheimer's drug results disappoint." Reisa Sperling was quoted in the *New York Times* as saying, "I don't know that we've hit a home run yet. It's important not to overconclude on the data. But as a proof of concept, I feel like this is very encouraging."

A shock to the Alzheimer's community occurred on March 21, 2019 when Biogen announced they were stopping their Phase 3 trials of aducanumab. Interim analysis showed that it was futile to continue the trial. Based on the results thus far, there was no chance that it could work. Al Sandrock's one-million-dollar investment did not pay off. He felt devastated, but he had been there before when Biogen's ALS trial had failed and when the MS drug Tysabri was taken off the market within months of its approval because it caused fatal brain infections.

In addition to the trial of Eisai's BAN2401, there are two other antibodies against A-beta that are in late-stage development by Genentech/Roche, one called crenezumab, the other called gantenerumab. Two independent Phase 3 trials are underway for crenezumab, which binds A-beta oligomers, the most toxic form of A-beta. Dennis and coworkers at our Center and at Genentech demonstrated for the first time that crenezumab showed target engagement by reducing A-beta oligomers in the spinal fluid of treated patients. Two Phase 3 trials have also been initiated by Roche for gantenerumab, which binds to aggregated beta-amyloid. Another anti-A-beta antibody called donanemab being developed by Eli Lilly showed positive results in a small two-year study involving 272 patients, as measured by slowing cognitive decline and clearing A-beta from the brain.

The big question, of course, is given all the failures, will Eisai's antibody or Genentech/Roche's two anti-amyloid antibodies or Eli Lilly's antibody ultimately work? Analysts write that these companies are "in search of positive data in a field littered with the wreckage of earlier clinical crashes."

Dennis Selkoe was approached to address the question of the failures and the path forward. In an invited commentary for *Nature Reviews Neurology*, Dennis posited four potential explanations for the failed aducanumab trial. First, the amyloid hypothesis is wrong. This seems unlikely, given the presence of amyloid at the crime scene and the overwhelming evidence that has accumulated over the past thirty years pointing to a critical role for amyloid. Second, the patients were treated too late in their disease. We know that the crime scene becomes very complex once the amyloid is deposited and clearing amyloid at this time wouldn't work. If this is the case, prevention trials in which anti-amyloid antibody is given to asymptomatic subjects should work. Third, it may be that the anti-amyloid antibody is not strong enough to clear enough amyloid to have an effect, or it is binding to a nontoxic form of amyloid. Finally, it may be that combination therapy that targets not only amyloid but other parts of the crime scene as well is needed. For example, it often takes multiple drugs to control high blood pressure or treat cancer.

Thus, the challenge for the treatment of Alzheimer's disease is to not focus solely on anti-amyloid therapy but to find treatments that can target tau, or neuronal dysfunction, or inflammation, which are all downstream consequences of the earliest accumulation of toxic A-beta-42, which triggers the entire process.

Then, on October 22, 2019, seven months after Biogen announced that aducanumab had failed futility analysis, Biogen shocked the medical community again when it announced that additional analysis of the ENGAGE and EMERGE trials of aducanumab, with over 1,300 patients in each trial, was in fact not negative. Reanalysis of the data that included additional patients receiving the higher dose, yielded significant positive results in those treated with higher doses of the drug versus placebo for longer periods of time in the EMERGE study. Two months later, at a much-anticipated scientific presentation, Biogen shared their findings at the 12th Clinical Trials in Alzheimer's Disease Conference held on December 4–7 in San Diego.

What had changed in the seven months from the futility announcement, and what did they report in San Diego? Evidently, at the time of futility analysis in March 2020, EMERGE was trending positive, whereas ENGAGE was not. Then, the EMERGE trial became positive when additional patients were followed at the high dose. Although ENGAGE remained negative, in a subsequent analysis, data from a subset of patients given high-dose aducanumab in ENGAGE supported the positive findings of EMERGE. Furthermore, in sub-studies aducanumab showed an effect on disease-related biomarkers.

A commentary in *Lancet Neurology* in February 2020 entitled "A Resurrection of Aducanumab for Alzheimer's Disease" highlighted potential pitfalls in the Biogen analysis and raised questions about drug approval. Nonetheless, Biogen submitted aducanumab for approval to the FDA with the hope of marketing and treating selected Alzheimer's patients. Furthermore, in March 2020, in the continued development of aducanumab, Biogen launched a re-dosing trial in patients who were previously enrolled in both the EMERGE and ENGAGE trial, in which patients will receive monthly doses for one hundred weeks with the primary outcome being safety and imaging abnormalities.

In August 2020, the FDA accepted Biogen's aducanumab Biologics License Application (BLA) with priority review and promised a decision by March 7, 2021. In addition, the FDA said they planned to hold a scientific advisory committee meeting to review the application in the fall of 2020. FDA scientific advisory committees are convened when there is uncertainty or controversy over whether a drug should be approved. The recommendation of an advisory committee is not binding, although the FDA usually, but not always, acts in accordance with the scientific advisory committee's recommendations. I have served as an expert on FDA scientific advisory committees for MS drugs and have traveled to Washington, D.C., to observe MS advisory committee meetings in person. At scientific advisory committee meetings, the FDA advisory panel sits at a table in the middle of the room, with representatives of the FDA on one side and representatives of the company on the other. Scientific advisory committee meetings are open to the public. In some ways, it is like a courtroom with high drama when the scientific advisors vote in front of everyone to recommend approval or not. Dennis had planned to travel to the FDA meeting, but because of COVID the scientific advisory committee meeting was held virtually. The live webcam of the meeting was so popular that some could not connect live and had to watch the recorded version later.

Whether aducanumab would or should be approved by the FDA was hotly debated both in scientific and investor circles. If approved, it would become the first disease-modifying therapy for the treatment of Alzheimer's and a major milestone that would energize the entire field. Furthermore, it would be a financial bonanza for Biogen.

A scientific advisory committee meeting was scheduled for Friday, November 6, 2020. On the Wednesday prior to the meeting, Biogen's stock jumped 42 percent when the FDA said on its website that the data presented to the FDA were "highly persuasive" and that "the application has provided substantial evidence of effectiveness to support approval." In addition, the FDA said there was "an acceptable safety profile that would support use in individuals with Alzheimer's disease."

The advisory committee meeting began its six-hour virtual meeting at 10 AM on Friday, November 6. In a prerecorded presentation, Al Sandrock introduced himself as a neurologist, neurobiologist, and head of research and development at Biogen. He said that fifteen years earlier, Biogen had intensified its effort in Alzheimer's disease, which led to studies of aducanumab, a monoclonal antibody that specifically targeted aggregated forms of amyloid-beta. Biogen opened an IND for aducanumab in 2011 and conducted Phase 1 trials, which showed that aducanumab cleared amyloid-beta in a dose-dependent fashion and was also associated with positive clinical effects. This led to a Phase 3 trial in 2015 under a Special Protocol Assessment with the FDA, and in 2016 fast-track status for aducanumab was granted by the FDA. Sandrock then reported on the futility analysis that led to termination of the Phase 3 trials in March 2019, only to discover a few months later that one of the trials was apparently positive. They shared the data with the FDA, who agreed that there was evidence of efficacy. That led to the filing of a BLA in July 2020.

Dr. Samantha Budd Haeberlein, head of the Neurodegenerative Development Unit at Biogen, then presented. She reviewed the three studies that had been carried out with aducanumab. She began with the EMERGE study, which, she argued, was positive with robust internally consistent results. She then discussed the failed ENGAGE study and pointed out that a subset of patients who received the appropriate dose were indeed positive and why it failed when the EMERGE study succeeded. Finally, she discussed the Phase 1 proof-of-concept PRIME study published in *Nature,* which led Biogen to launch the EMERGE and ENGAGE studies. She argued that there was a consistent effect of lowering amyloid-beta at the 10mg/kg dose across the studies and that positive clinical effects were seen. She said that the failed ENGAGE study should not detract from the persuasiveness of the EMERGE study.

Safety data were then presented by Dr. Karen Smirnakis, head of global medical safety at Biogen. The main adverse effect was amyloid-related imaging abnormalities, or ARIA, consisting of brain swelling and brain micro-hemorrhages as a result of aducanumab binding to the amyloid-beta in the brain. Although more common in treated patients, especially at the highest 10mg/kg dose, she pointed out that most patients did not experience ARIA or had mild symptoms and were able to complete treatment. Furthermore, the risk could be mitigated with routine MRI monitoring and dose management.

Dr. Billy Dunn, director of the FDA Office of Neuroscience, gave the FDA's perspective on the trials. He said that the positive EMERGE study was highly persuasive and provided substantial evidence of the effectiveness of aducanumab. He said that the FDA's analysis had concluded that the negative ENGAGE study did not contradict the EMERGE study or show that aducanumab was ineffective. Furthermore, the drug had an acceptable safety profile.

However, as the meeting proceeded, biostatisticians raised concern. A prerecorded statistical review of aducanumab by Dr. Tristan Massie, a statistician from the Division of Biometrics at the FDA, pointed to flaws in the analysis, emphasizing that there was a lack of substantial evidence with no replication, highly conflicting results in the two studies, and conflicting subgroup analysis. He was also concerned that if approved, it would create recruitment challenges for ongoing trials and actually impede the development of an effective disease-modifying drug for Alzheimer's. He concluded that there was no compelling, substantial evidence of a treatment effect or disease slowing and that another study was needed. At the advisory board hearing, Dr. Scott Emerson, a professor emeritus of biostatistics at the University of Washington in Seattle, echoed his concerns and poured cold water on the aducanumab results. He did not like the idea of ignoring the negative study, which he likened to choosing two numbers and then revealing only the highest number. He was disturbed that the FDA seemed to be starting out with the assumption that adacanumab worked and then trying to figure out why the other trial failed, which he felt was a one-sided view of the results.

After a lunch break, there was an open public hearing in which sixteen people spoke. All but two urged the committee to approve aducanumab. The speakers told poignant stories of themselves or family members suffering from the disease. George Vradenburg, chairman of the UsAgainstAlzheimer's nonprofit organization, argued that in order to get the best-in-class drug, one needed a first-in-class drug, and that the signal sent by the FDA could either stop the path toward a cure or be the starting gun on innovation. Joanne Pike, chief strategy officer of the Alzheimer's Association, argued against the need for an additional Phase 3 study, which would deny access to a treatment for Alzheimer's disease for another four years. However, Dr. Joel Perlmutter, a panel member and professor of neurology at the Washington University School of Medicine, argued that aducanumab did not appear effective, and if approved, patients would not participate in other trials, and an effective treatment would be delayed for many years. Another panel member, Dr. Caleb Alexander, a professor of epidemiology and medicine at Johns Hopkins, said there were a dozen different red flags that suggested concern about the drug's efficacy.

Finally, the vote came. Ten of the eleven advisory board members voted against approving aducanumab. They felt that the research presented did not provide "primary evidence of the effectiveness of aducanumab for the treatment of Alzheimer's disease." Many in the Alzheimer's community were shocked and disappointed. Dennis Selkoe wrote in the AlzForum that the FDA's decision was a clear setback. He felt that there had been insufficient time to address the advisory committee's questions about the discrepancies between the positive and negative trials, which might have simply been related to differences in exposure. He suggested that all data should be made publicly available, blood biomarkers from the studies analyzed, and emerging data from the ongoing EMBARK open-label extension trial be considered by the FDA. He cautioned against setting aside an agent that had been tested in thousands of patients and that appeared to confer some benefit in a subset of patients with manageable safety.

On the Monday after the advisory board meeting, Biogen's stock fell as much as 32 percent in premarket trading, erasing most of the gains that it made the previous week. Al Sandrock told me he was less bothered by the fall in Biogen's stock than the possibility that a drug he truly believed Biogen had shown to help Alzheimer's disease would not be made available to patients. He contrasted that to his feelings when a Phase 3 trial Biogen drug for ALS called dexpramipexole failed. That drug had simply not worked. Aducanumab did.

The FDA is not bound by the advisory committee's opinion. Thus, AD physicians and patients had to wait for the FDA to release its verdict. Initially scheduled for March 7, 2021, the FDA delayed its decision until June 7, 2021. Some felt it was a good sign. Why would the FDA delay a decision if they had decided to reject aducanumab?

In the interim, on March 13, 2021, the Alzheimer's community was greeted with exciting news when Phase 2 results of donanemab, Eli Lilly's anti-A-beta antibody, were published in the *New England Journal of Medicine* and presented at the 15th International Conference on Alzheimer's and Parkinson's Diseases (held virtually because of the COVID pandemic). Lilly's TRAILBLAZER-ALZ trial randomized 131 early symptomatic Alzheimer's patients to receive donanemab or placebo intravenously every month for up to 72 weeks. Investigators had to screen 1955 patients to find 257 that met eligibility requirements. The primary outcome measure was improvement on the Integrated Alzheimer's Disease Rating Scale, a composite measure of cognition and daily function. The study met its primary outcome measure as there was significantly less cognitive decline in those receiving donanemab than in those receiving the placebo. Importantly, brain imaging showed clearance of amyloid in the treated group. A pivotal trial called TRAILBLAZER-ALZ2 is underway.

The Alzheimer's community anxiously awaited the FDA's June 7th decision and was split on whether aducanumab would be approved. I polled the scientists during my Monday morning lab meeting: half predicted approval, half rejection. Dennis Selkoe expected the FDA announcement in the afternoon after the stock market closed. The news came before noon. Aducanumab was approved and would be marketed under the trade name Aduhelm making it the first disease-modifying therapy for Alzheimer's disease and the first Alzheimer drug approved since 2003. Biogen stock rose over 30 percent when the news broke.

Aduhelm was met with enthusiasm by the Alzheimer community but with controversy by some in the scientific community. Three members of the FDA Advisory Committee resigned in protest since they recommended the drug not be approved. Aduhelm was approved under the Accelerated Approval pathway and Patricia Cavazzoni, Director of the FDA Center for Drug Evaluation explained the FDA's decision on the FDA website. She acknowledged that the data were complex, left uncertainties regarding the drug's clinical benefit, and that the expert committee had a different perspective. She also acknowledged that the option of Accelerated Approval was not discussed by the Advisory Committee. Nonetheless, the Accelerated Approval pathway was intended to provide earlier access to potentially valuable therapies for serious diseases even if there was uncertainty regarding clinical benefit. Although only one of Biogen's studies had shown clinical benefit, Aduhelm consistently and convincingly reduced amyloid in the brain in a dose and time-dependent fashion. The FDA expected that reduction in amyloid would lessen clinical decline analogous to reducing heart attacks by lowering blood pressure. Moreover, Biogen was now required to conduct another study to verify the clinical benefit and the FDA could remove the drug if the confirmatory trial was negative. Despite the imperfect clinical data, Cavazzoni emphasized that Alzheimer's was a devastating disease and that the FDA felt approval of Aduhlem would provide a critical new treatment to combat the disease.

On the Friday after Aduhelm's approval, Al Sandrock from Biogen spoke at our hospital. He reminisced about his time at Mass General and Harvard Medical School before going to Biogen and recalled a seminar in the 1980s when he heard Dennis speak about amyloid and Alzheimer's disease, at a time when Alzheimer's was not well recognized. He reviewed the development of aducanumab from its first being licensed from Neurimmune in 2007 and how it differed from other anti-A-beta antibodies. He said that Aduhelm also affected tau in the brain and that its action likely involved microglia.

Aduhelm was broadly approved for Alzheimer's, though it was later changed for only mild disease. It is given by monthly infusion in an escalating dose to minimize ARIA (amyloid related imaging abnormalities including brain swelling) which would require MRI monitoring. Headache was the most common symptom patients experienced. Some thought that Biogen's $56,000/year price tag was too high. It is unfortunate that the first disease-modifying treatment for Alzheimer's disease was ushered in amidst controversy, but there was no doubt it was a historic milestone that would energize the field. In a commentary on Aduhelm for *Science*, Dennis wrote that in therapeutics as in life, one must walk before one runs.

On June 21, the Alzheimer's Association hosted an expert forum attended by 1400 participants. All agreed Aduhelm should be limited to mild AD with proven brain amyloid. "Oy," said Dr. Stephen Salloway, an Aduhelm supporter, in describing his reaction to no contraindications in the label. Important questions remained: how to measure efficacy in individual patients and when to stop the drug. Not all doctors have the expertise to use Aduhelm and it is crucial to manage patient expectations. Nonetheless, all agreed Aduhelm was a win-win for both doctors and patients.

PREVENTING ALZHEIMER'S

The ultimate treatment for Alzheimer's, of course, must rest with prevention and early intervention, which requires deeply understanding the preclinical aspects of the disease where symptoms have not yet appeared. It has now become clear that there is a long preclinical stage of Alzheimer's disease in clinically asymptomatic people. This may, indeed, be the best time to begin therapy, but a major question is how to identify preclinical Alzheimer's disease. CSF and blood biomarkers are being studied, and with the advent of amyloid brain imaging in 2002, amyloid deposits can be measured.

The first preclinical accumulation of beta-amyloid plaques measurable by amyloid imaging probably occurs ten to twenty years before any symptoms. This is followed by synaptic dysfunction of neurons and then tau-mediated neuronal injury, which can be measured both in spinal fluid and by tau brain imaging. These features may be followed by the initial changes in brain structure as measured by MRI. Around this time, there may be mild cognitive impairment. In a few years, most patients begin to show frank dementia—that is, substantial clinical dysfunction. It is possible that preclinical neuronal injury may be detectable earlier in people who carry the *ApoE4* gene, presumably because they have abnormal clearance of toxic A-beta-42 for many years.

There are at least three preclinical (presymptomatic) AD trials underway, all using an antibody against amyloid: Reisa Sperling's A4 trial discussed previously, a trial centered in Medellín, Colombia, and trials being conducted by the Dominantly Inherited Alzheimer's Disease Network (DIAN). The Colombian subjects and those in the DIAN study have mutations in the presenilin genes, which cause a dominantly inherited form of Alzheimer's—everyone who carries the mutant gene gets the disease. This was the type of AD that the actress Julianne Moore depicted in the movie *Still Alice*. The trial in Colombia is funded by Genentech, the National Institute of Aging, and the Banner Institute in Phoenix. Genentech's anti-amyloid antibody crenezumab is being administered. It is a double-blind study in which one hundred asymptomatic carriers of a presenilin mutation will receive the drug, one hundred will receive placebo, and one hundred non-carriers in the same age group will also receive a placebo. If these presymptomatic trials don't work, they will demonstrate that either the amyloid hypothesis is somehow wrong, or the anti-amyloid antibodies being used are not able to clear enough amyloid from the brain and neutralize the toxic effects of A-beta.

Other trials to prevent Alzheimer's are also being conducted. It is known that people with type 2 diabetes are more likely to get dementia, including Alzheimer's disease. An analysis of hundreds of thousands of type 2 diabetes patients found something striking. Those who took the diabetes drug pioglitazone were less likely to develop Alzheimer's disease. A trial called the TOMORROW study treated almost six thousand cognitively normal patients with pioglitazone to see whether it delayed the onset of Alzheimer's. Unfortunately, the results were negative. Pioglitazone has also been tested in patients with ALS without benefit. Other preventive therapies have also been tried and failed, including ginkgo biloba and hormone therapies. However, a randomized controlled trial testing lifestyle modifications that include diet, exercise, cognitive training, and decreasing risk of vascular disease is being carried out in Finland. It is called the FINGER trial (Finnish Geriatric Intervention Study to Prevent Cognitive Impairment and Disability) and has shown positive results in preventing cognitive decline. This has led to a similar study in the United States called the POINTER trial (Protect Through a Lifestyle Intervention to Reduce Risk). Reisa Sperling has the A3 trial (Ante-Amyloid prevention of Alzheimer's disease) in clinically normal older individuals without elevated amyloid on PET imaging.

With our understanding that Alzheimer's begins years before there are any symptoms, biomarkers become crucial, including those that identify people at risk of developing dementia who do not have Alzheimer's disease. Using markers of neurodegeneration that included spinal fluid exams and imaging, patients designated as having SNAP (Suspected Non-Alzheimer disease Pathophysiology) were followed at Washington University in St. Louis and as part of the Harvard Aging Brain Study and were found not to develop Alzheimer's disease. These people have neuronal changes, including abnormal tau accumulation, in brain regions important for cognition but with no A-beta deposits.

Clearly, understanding the different subtypes of dementia will be crucial in understanding why certain therapies work or do not work, especially those targeting A-beta. In the CSF, there's a reciprocal relationship between A-beta-42 protein (which is lower in AD) and tau (which is higher in AD).

Recently, blood biomarkers have emerged that are specific for AD and can be used as an early biomarker.

A protein called neurofilament light (NfL) is released from damaged neurons and enters the blood. The marker appears to differentiate between healthy control and Alzheimer's patients, but because changes in NfL levels are also found in multiple sclerosis, frontotemporal dementia, and other progressive neurological diseases, it is not specific for Alzheimer's but is a general marker of neuronal injury in the brain.

On the other hand, subunits of certain forms of tau in the blood (NT1 tau and pT217 tau) appear to be a specific marker for AD, which is elevated early in disease and could prove useful as a first round to screen and identify individuals at risk for developing AD. Studies by Dennis Selkoe's group found that plasma levels of the NT1 fragment of tau taken when people were still normal cognitively but had low levels of amyloid plaques (as seen on brain PET imaging) could predict the occurrence and even the rate of cognitive decline over the next six years. They could also predict the future occurrence of brain atrophy and accumulation of tau in tangles. Blood tests such as these are desperately needed, especially if we are to screen and monitor large numbers of the population at risk for Alzheimer's as disease-modifying therapies emerge.

Imaging the brain itself has been very helpful in understanding the disease process. It can demonstrate deposits of amyloid, deposits of tau, structural changes in parts of the brain affected by the disease, and microglial activation. The two major features at the crime scene of Alzheimer's are plaques and tangles. Amyloid protein accumulation comes first, which then triggers tau protein accumulation. Tau accumulation is linked to disease severity and dementia, which contributes to cellular damage. Interestingly, genes whose proteins are necessary for amyloid accumulation are strongly associated with Alzheimer's, but genes whose proteins are necessary for tau accumulation have never been found to be abnormal in an Alzheimer's patient. This suggests that tau is not the primary cause of Alzheimer's. Mutations in only the tau gene cause a different type of dementia, called frontotemporal dementia, which is also referred to as tauopathy. Tauopathies affect different brain regions, progress at different rates, and have different biochemical patterns of tau accumulation from what is seen in Alzheimer's disease.

TARGETING TAU

Given the many failures of trials targeting the A-beta pathway, a question arises: Why not attack Alzheimer's by developing therapies directed against tau in the same way attempts have been made to develop therapies directed against amyloid? This approach is indeed underway both by giving antibodies against tau and by immunizing with tau. The drug company Boehringer Ingelheim has developed interest in compounds that inhibit tau aggregation and enhance stabilization of tau in the cell by affecting microtubules. Just as the development of a mouse model of amyloid in the brain was used to test therapies, mouse models of tau disease are being used. In addition, in the same way that brain PET imaging of amyloid has allowed us to view the amyloid and how it relates to Alzheimer's disease, the same is now true for brain PET imaging of tau, which has turned out to be an indicator of the progression of clinical symptoms and neurodegeneration in Alzheimer's disease. Different patterns of imaging abnormalities are found when comparing amyloid with tau brain imaging. Studies have long shown a better correlation between degree of dementia with tau tangle load than with beta-amyloid burden.

Tau is located predominantly within the cell, whereas amyloid-beta deposits are outside the cell, thus making tau more difficult to target. Nonetheless, small amounts of tau are released from neurons. Studies in tau transgenic mice have shown that administering an antibody against tau can reduce the disease in the brain, improve function, and, perhaps, block spread of tau from cell to cell. Trials in mice using anti-tau antibodies showed increased tau levels in the plasma in such mice, which express human tau. The same has been shown when antibody against tau was given to patients with progressive supranuclear palsy, a tau-related brain degenerative disease. The antibody is believed to bind some of the tau that comes out of the cell. Trials are underway to treat Alzheimer's disease patients with antibodies to tau in the same way that antibodies to amyloid are given. Because tau appears after amyloid in Alzheimer's disease and is more closely linked to cognitive loss than amyloid, antibodies to tau may be effective in Alzheimer's disease, where antibodies to amyloid have failed.

Another approach that has been used to reduce tau in mouse models is antisense oligonucleotides, or ASOs. ASOs are a form of gene therapy in which the RNA made by a specific gene is silenced. Anti-tau ASO gene therapy decreases tau in these mice and results in less neuronal death and longer survival. As we will see in the next chapter, the use of ASO therapy to silence bad genes is being tried in ALS, and it has had a dramatic effect in a lethal muscle disease in children called spinal muscular atrophy, for which the ASO has received FDA approval.

Electrical activity in the brain known as gamma brain waves helps connect and process information throughout the brain; these gamma waves are diminished in Alzheimer's disease. In 2016 Li-Huei Tsai of the Picower Institute for Learning and Memory at MIT published an article in *Nature* reporting that treatment of Alzheimer's mice with a light flickering at the rate of forty flashes per second (40Hz) increased gamma brain waves, decreased amyloid load in the brain, and modified microglia. An observational five-year clinical trial in two thousand patients began in 2018 in which subjects will undergo 40Hz light therapy using a novel iPad app called ALZLIFE plus cognitive therapy.

AMYLOID VERSUS TAU

In the past, there has been a debate about the role of amyloid versus tau in Alzheimer's disease. It is now accepted that both play an important role, with amyloid protein accumulation coming first and then triggering tau accumulation.

One of the factors linked to Alzheimer's disease that affects multiple pathways, including both amyloid and tau, is *ApoE4*. As discussed, the E4 form of *ApoE* is a genetic risk factor for late-onset Alzheimer's disease, and approximately 50 percent of people with Alzheimer's carry *ApoE4*. *ApoE4* is such an important factor that in clinical trials patients are often segregated into *ApoE4* positive or *ApoE4* negative groups.

In animal models, *ApoE* can act on both A-beta and tau. In A-beta models, *ApoE* facilitates deposition and accumulation of A-beta because it is an A-beta-binding molecule. In the tau model of disease, *ApoE* worsens the neurodegeneration independent of its effect on A-beta. Furthermore, microglia that express the E4 form of *ApoE* have a neurodegenerative disease-inducing Mr. Hyde phenotype. Thus, *ApoE4* affects the triad of actors at the crime scene: A-beta, tau, and microglia. *ApoE4* is an obvious drug target in Alzheimer's disease, and investigators are determining whether drugs that alter the shape of or knock out the *ApoE4* gene could be beneficial in Alzheimer's disease.

IMMUNITY, THE GUT, INFECTION, AND ALZHEIMER'S

Interestingly, although both amyloid and tau play a role in the disease, they can be affected differently by the immune system. Alzheimer animals that have excess A-beta in the brain, and that are deficient in an immune protein on the surface of microglial cells, have enhanced clearance of A-beta. However, the absence of the immune protein on microglial cells has no effect on tau in animals with excess tau in the brain. These studies emphasize how complicated the immune system's role is in Alzheimer's disease and that the same immune treatment could make Alzheimer's disease better or worse depending on the stage of the disease when it is given.

The immune system's role in Alzheimer's disease is both beneficial and detrimental, making it an attractive target for drug therapy. At one time, it was thought that anti-inflammatory drugs such as ibuprofen could help Alzheimer's, presumably by decreasing abnormal inflammation in the brain. People who took anti-inflammatory drugs appeared to have a decreased incidence of Alzheimer's disease. However, a large trial was not positive. On the downside, there have been reports that cladribine, a drug used for MS that suppresses the immune system, leads to more accumulation of A-beta in the brain in mouse models.

Other immune-directed approaches that have had positive clinical results in MS and ALS may also benefit those with Alzheimer's disease. An example is masitinib, an orally administered drug that targets microglial cells and mast cells. In a six-month Phase 2/3 trial of 718 patients with mild to moderate Alzheimer's disease, masitinib halted cognitive decline and showed improvement of the ADAS-Cog, the widely used Alzheimer's clinical disease scale.

These results suggest that although the immune system may not be the primary driver of brain disease, the immune system plays an important role in amplifying and contributing to disease progression and thus is a good therapeutic target for more than one brain disease. We know that microglial cells in the brain can become toxic and contribute to brain damage in many diseases. As discussed in the chapter on MS, we have found that nasally administered anti-CD3 monoclonal antibody dampens microglia inflammation and ameliorates disease in an animal model of progressive MS. Based on this, we are initiating a clinical trial of nasal anti-CD3 in progressive MS, and we have found that nasal anti-CD3 also ameliorates disease in models of Alzheimer's disease and ALS.

Rudi Tanzi, who heads the Genetics and Aging Research Unit at Massachusetts General Hospital, told me how a drug called cromolyn, commonly used to treat asthma, came to be tested in Alzheimer's disease. Cromolyn was initially investigated for its potential to block amyloid formation in the test tube, only to be found to reduce amyloid in mouse models. It appears to work by stimulating the microglia in the brain to remove amyloid. Cromolyn, which is given by inhalation, is now in a Phase 3, randomized, placebo-controlled trial for subjects with early AD. Rudi Tanzi told me that some of the patients in the trial seem to be benefitting, but because the trial is double blind, he doesn't know who's getting the medicine and who's not.

As discussed in the chapter on multiple sclerosis, the gut microbiome plays a key role in many diseases, including those of the brain. This is also true for Alzheimer's disease. As people get older, the gut microbiome changes. In fact, it has been shown that the composition of the gut microbiome correlates with diet and health in the elderly. There are studies linking obesity and other metabolic disorders with an increased risk of Alzheimer's disease, and this, in turn, may also be linked to the gut microbiome. As people get older, changes occur in the permeability of the intestine, and it becomes more leaky. This could result in the release of bacteria or bacterial products into the bloodstream, which could in turn affect the immune system and the brain. As we get older, we develop a frail gut that may in turn contribute to a frail brain.

A number of observations both in animal models of Alzheimer's and in patients with Alzheimer's disease suggest a link to the gut microbiome. This connection relates primarily to the inflammatory component of Alzheimer's and the effects of the gut microbiome on microglial cells and monocytes. The gut microbiome is necessary for normal functioning. Remarkably, the microglia in germ-free mice that do not have a normal gut microbiome are not as healthy, taking on a Mr. Hyde appearance. Thus, if you perturb the gut and eliminate most of the microbes, how does this affect the disease in animal models of Alzheimer's? A series of three experiments shows that the microbiome has a positive effect and raises the possibility that an ultimate treatment of Alzheimer's may include manipulation of the gut to affect the gut-brain axis.

In the first series of experiments, investigators perturbed the gut in animal models by giving broad-spectrum antibiotics. This treatment kills most of the bacteria in the gut and changes the composition of the microbiome. They found that Alzheimer's mice treated with the antibiotics had alterations in the microbiome and less inflammation in the blood. Mice that received antibiotics also had decreased amyloid deposits in the brain, and there was reduced scarring surrounding the plaques in these mice. The treatment affected serum levels of a chemical called CCL11, which can cross the blood-brain barrier and lead to activation of microglia and clearance of A-beta. Interestingly, these changes were seen primarily in male, as opposed to female, mice. The reason for this difference is unknown.

In a second series of experiments, investigators performed gene analysis of all the bacteria in fecal samples from Alzheimer's transgenic mice and compared them to the bacteria in normal mice. They found a difference in the microbiome of the Alzheimer's transgenic mice. Then, in an experiment similar to treating mice with antibiotics, they generated germ-free transgenic mice in which there were no bacteria in the gut. In mice with no gut bacteria, there was a drastic reduction of amyloid in the brain. They then conducted a transfer experiment in which they colonized germ-free Alzheimer's mice with feces from transgenic mice that had a microbiome that wasn't altered and found an increase in the signs of the disease. They were able to identify a specific bacterium that was linked to influencing the disease. Thus, Alzheimer's mice can have a different microbiome from normal mice, and there is something in the gut microbiome of Alzheimer's mice that makes the disease worse.

In a third series of experiments, investigators studied the effect of calorie restriction in Alzheimer's mice. Calorie restriction has been shown to reduce Alzheimer's pathology both in mice and in monkeys. It appears to work through multiple mechanisms, including modulation of the function of the gamma-secretase scissors that cuts the amyloid protein to produce toxic A-beta-42, and increases the scavenger function of immune macrophages. Interestingly, some of the effects of calorie restriction in Alzheimer's disease are seen only in female mice. In our studies, we found that the protective effect of calorie restriction in female mice was related to the microbiome and to the bacteria *Bacteroides fragilis*. Increased *Bacteroides fragilis* was linked to increased A-beta plaques, and this was reversed by calorie restriction. Conversely, giving *Bacteroides fragilis* to mice on a normal diet increased A-beta plaques. Others have found that *Bacteroides fragilis* increases with aging and is increased in Alzheimer's disease.

A controlled study at the Harvard School of Public Health called the MIND trial is testing whether a weight loss diet can affect brain health and cognitive decline in people who have a family history of Alzheimer's disease. Initial studies suggest that there are changes in the microbiome in patients with Alzheimer's disease. In fact, similar changes have been observed in patients in some of our calorie restriction studies.

It is likely that factors released from the gut directly affect the brain and the development and progression of Alzheimer's. It is also likely that changes in the immune system, both with microglial cells and monocytes, may be related to the gut microbiome and could also affect the disease. We found that treating Alzheimer's mice with antibiotics could actually change microglia in the brain and convert the good Dr. Jekyll type to the bad Mr. Hyde type. The study of the microbiome in Alzheimer's disease is in its infancy but will grow exponentially in the coming years, as it will for the other neurological diseases we discuss in this book.

One of the major causes of disease that has been successfully treated is infectious agents, such as polio or smallpox virus. An intriguing hypothesis related to Alzheimer's disease is that infection plays an important role in the disease, encompassing viruses such as herpes virus, a type of *Chlamydophila* that causes pneumonia, and even spirochetes, which are related to syphilis and Lyme disease. A recent study looking at network analysis of the brain reported that human herpes virus may be linked to the disease, and other investigators suggest that amyloid-beta may protect against fungal and bacterial infections and that developing amyloid plaques in the brain might in some instances be a beneficial response to infection. An organization called Alzheimer's Germ Quest is lobbying for more support for this hypothesis and offering a prize for "persuasive proof" that the "Alzheimer's germ" is the cause of the disease. Many researchers in the field are skeptical. There were many reports of an "MS virus," which could never be found, although infection with *H. pylori* was unexpectedly shown to cause ulcers. The ultimate proof, of course, will rest on identifying specific organisms and demonstrating that Alzheimer's disease can be treated by targeting them in some way.

PREVENTION AND CURE?

People always ask what can be done to protect against Alzheimer's. Apart from choosing your parents wisely, it appears that keeping socially and intellectually active, watching your diet, and exercising can help stave off Alzheimer's. In terms of keeping intellectually active, I worked with Dennis on an experiment called Environmental Enrichment in mice. Mice are usually housed in simple cages with food and water. Mice in environmental enrichment cages had toys placed in their cages. As it turned out, mice with toys in their cages developed less amyloid in the brain, and the environmental enrichment actually affected microglia in the brain, making the microglia act like Dr. Jekyll as opposed to Mr. Hyde. Reisa Sperling also has accumulated evidence that getting enough sleep can help protect the brain against Alzheimer's disease.

How then will we cure Alzheimer's disease? As it is with MS, it depends on the definition of a cure. The ultimate cure would be to prevent Alzheimer's so that it never happens. In the future, this could theoretically occur by genetic manipulation. There is a genetic mutation called the Icelandic mutation that protects people against Alzheimer's disease, and people with this mutation make substantially less amyloid-beta and never get AD. The mutation was discovered in the Icelandic population, which served as a living laboratory for the investigation of the genetic links to disease, given that Iceland was isolated and the population inbred for so many years. The Icelandic mutation is in APP, which, of course, fits directly with the amyloid hypothesis, as this genetic mutation decreases the production of A-beta. Could we introduce the Icelandic mutation into the general population? Such an approach is far in the future, but understanding genetics and the ability to affect genes could be the ultimate answer to neurological and other diseases where causative or protective genes are identified.

We have now identified several modifiable factors that contribute to Alzheimer's disease: A-beta, tau, the immune system, and lifestyle. We also have a better understanding of the chronology of events that leads to clinical Alzheimer's disease. Anti-A-beta strategies are most likely to be effective early in the clinical disease or, better yet, in prevention trials, and less effective in later stages of the disease. In later stages, drugs that target tau, the immune system, or neurodegenerative processes may have a chance of helping. Ultimately, combination therapies will be employed, as they are in other chronic progressive diseases. Whether or not our nasal vaccine strategy works, treating large segments of the population at risk with an easily administered therapy will be needed. More than likely, people in their forties and fifties will be given something to strengthen the immune system's ability to deal with A-beta directly, as well as a compound to prevent the deposits of A-beta and thus the entire cascade.

Some approaches may also involve manipulation of the microbiome. Equally as important, environmental and lifestyle changes may go a long way to reducing the incidence of Alzheimer's. However, we are fighting an uphill battle, because we know that the older people get, the more they are at risk for developing Alzheimer's, with 25 percent of people over ninety having evidence of amyloid in the brain. In a recent talk, Reisa Sperling said that we don't have to permanently cure the disease; if we can delay it by ten years, many of those who would have contracted Alzheimer's will die from other causes while attending their ballroom dancing classes.

I believe, as in MS, we now understand much more about the mechanisms causing Alzheimer's and their timing. Thus, we know which targets to attack and when. A major milestone was the FDA approval of Biogen's anti-amyloid drug Aduhelm. Three years after Pfizer's publication of their failed anti-amyloid study, they unfortunately laid off over three hundred people and stopped their research and clinical development in Alzheimer's disease. This, of course, is not the way to discover a solution. The approval of Aduhlem will serve as major impetus for industry to study the disease and provide a foundation to develop combination therapy and define responders and non-responders to therapy.

At the June 21 Alzheimer's Association's expert forum held just following Aduhelm's approval, Kathy Costello, an MS nurse who is Vice-President of Healthcare Access for the National MS Society, gave her perspective on Aduhlem. She remembered when Betaseron, the first FDA drug for MS, was approved in 1993. Like Aduhelm, it had only modest benefit in early disease and didn't help more advanced stages. Patients had to inject themselves and both side effects and expectations had to be managed. It was the first, but not the last drug for MS. There are now over 20 drugs for MS. Philip Scheltens, who is Director of the Alzheimer Center at Amsterdam University and who co-chaired the expert forum, said that Alzheimer's today is where the MS field was 20 years ago.

Crucial to our understanding and developing effective treatments for Alzheimer's disease is the ability to image the disease and the development of biomarkers. The first drug approved for MS occurred because of a breakthrough in MRI imaging that allowed doctors to visualize MS plaques and to show that treatment reduced them. For Alzheimer's, imaging for amyloid, tau, and microglia, as well as emerging blood biomarkers, enable us to perform trials in which we can demonstrate that these factors specifically affect the disease process. Furthermore, biomarkers will allow screening of large portions of the population, especially if they can be detected by a simple blood test.

It can be said without hyperbole that Alzheimer's disease is the disease of the century. I am optimistic that soon in this century it will become treatable. We have no other option. In fact, I believe there are people with Alzheimer's alive today who will become the first survivors of the disease.

CHAPTER FOUR

Amyotrophic Lateral Sclerosis

In the summer of 2016, I was driving home from a golf game with a doctor from our MS Center when I got a call from a neurologist at the University of Colorado Medical School, where I studied medicine. She had just finished examining my daughter-in-law's mother, Tanya, a woman in her seventies, who had complained of mild speech difficulty. I had been with her a few months earlier at a family event and hadn't noticed anything amiss. But Tanya, a pediatrician who grew up in Russia, had sensed something was wrong and told her daughter. A cardiologist found nothing wrong, so my daughter-in-law Liz suggested that her mother see a neurologist. An MRI of the brain was normal, but an electromyogram, or EMG, in which needles are placed in the muscles, showed an abnormal pattern of electrical activity consistent with a diagnosis of ALS. Liz and her father were not sure what the doctor meant when he mentioned ALS, but Tanya knew. She immediately broke down and left the room to cry. When I told my colleague from our MS Center about the call, all he could say was, "I'm so sorry."

I immediately brought Tanya, Liz, and her father, Lazar, to Boston to see Merit Cudkowicz for a second opinion at the Healey Center for ALS at Massachusetts General Hospital, our sister hospital. Merit is chair of neurology at Mass General and a world leader in ALS clinical trials. We are working together on how the immune system affects ALS and are planning to test new forms of immune therapy that we are developing at the Ann Romney Center. The doctor in Colorado who suspected that Tanya had ALS trained with Merit, and we assumed the diagnosis was correct.

Many people we see may already suspect they have ALS, but it is difficult for the patient and family when we confirm their suspicion. Understandably, both the patient and the doctor look for a way out of the diagnosis, and because there are usually further tests to perform, the final diagnosis may wait a few weeks. Tanya, Liz, and Lazar arrived in Boston for a final verdict.

ALS is amyotrophic lateral sclerosis. *Amyotrophic* refers to wasting of the muscles, which happens when the muscles no longer receive electrical impulses from the nerves. *Lateral sclerosis* refers to scarring that happens on the sides of the spinal cord where fibers that stimulate movement are found. Because ALS is a disease of motor neurons and not sensory neurons, sensation is preserved. Unlike Alzheimer's or Parkinson's disease, cognition is not usually impacted, and for many, muscles of the eyes and bladder also maintain their normal function.

ALS is also known as Lou Gehrig's disease and, in Europe, motor neuron disease. Lou Gehrig was a baseball player for the New York Yankees who held the record for the most consecutive baseball games played—2,130. The record stood for fifty-six years, a measure of endurance. It is ironic that an athlete of such strength and endurance contracted a disease that robbed him of his ability to move. He was diagnosed on June 19, 1939, at the age of thirty-six, when it became clear he was having trouble playing baseball. He was honored at a public ceremony a few weeks later on July 4, 1939 at Yankee Stadium, where he gave a famous farewell speech in which he said, "Fans, for the past two weeks you have been reading about a bad break I got. Yet today I consider myself the luckiest man on the face of the earth." It was more than a bad break. The disease progressed rapidly, and despite being treated with vitamin E shots, Lou Gehrig died on June 2, 1941. Since then, the disease has carried his name.

In Europe and the United States, the lifetime risk of ALS is about 1 in 400. Each year, approximately 2 per 100,000 people receive a diagnosis of ALS. About 10 percent are inherited, or familial, while the remaining 90 percent are sporadic, or occur with no family history. ALS typically strikes in the mid- to late fifties. Though rare overall, ALS is the most common neurodegenerative disorder of midlife. Men are almost twice as likely as women to be diagnosed with sporadic ALS. The ratio of men to women with familial ALS is closer to 1:1.

ALS is one of the more tragic of the neurological diseases. It is a diagnosis every doctor hates to make. Before diagnosing ALS, all other possible causes for the weakness or trouble with speech or swallowing must be ruled out. MRIs of the brain and spinal cord are performed to search for any major structural changes that could explain findings such as a tumor, a syrinx (a fluid-filled cavity in the spinal cord), or evidence of a stroke. Blood is screened for the presence of toxins or metabolism imbalances, and for abnormalities of the immune system. However, to an experienced neurologist, the diagnosis of ALS can usually be made after taking a history and performing a neurological exam that shows a specific pattern of muscle weakness, spasticity, increased reflexes, and fasciculations, or the presence of twitching of the muscles in the limbs or the tongue. The diagnosis is confirmed by electrical testing in which needles are placed into the muscle to record electrical impulses. The muscles show characteristic changes of ALS in that they do not receive normal nerve impulses.

Now and then I receive a call from a frightened medical student who notices fasciculations or twitching of the calf muscles and tells me in hushed tones that they are afraid they have ALS. Invariably, it is a benign event. As a student, I, too, worried about fasciculations in my calf and, on occasion, I looked carefully at my tongue to make sure there was no twitching there. Despite these fears, ALS is rare enough in the general population to be considered an orphan disease, a condition affecting fewer than 200,000 people.

I sat with Tanya, Liz, and Lazar in an exam room at Mass General. We talked about our shared grandchildren as we waited for Merit to appear. I was usually the one to arrive a few minutes late when I saw MS patients at the Brigham. Now I was on the other side. Merit arrived with a comforting smile and took a seat next to Tanya. I moved to a corner so as not to interfere. Merit reviewed all the records from the doctor in Colorado and then performed a careful neurological examination, including arm and leg strength, reflexes, and strength around the mouth. Her arms and legs were strong, although her reflexes were increased. She had Tanya puff out her cheeks and push her tongue against Merit's hand. A slight weakness was noted. Merit then examined Tanya's tongue with her flashlight. Merit later told me that she saw tongue fasciculations.

When the exam was over, Tanya asked directly, "Do I have ALS?" Merit responded with a simple "Yes," immediately followed by "but there is a lot we can do." Tanya's eyes teared up and Merit gave her a hug.

There are two forms of ALS, which are determined by how the disease begins: limb onset and bulbar onset. In limb onset ALS the first sign is in one of the limbs, such as having trouble buttoning a shirt or difficulty walking or running. In bulbar ALS the brain stem is affected, and the first symptoms are slurred speech or difficulty swallowing. The bulbar form has a poorer prognosis. Tanya had the bulbar form of ALS.

Although the mean survival time with ALS is three to five years from diagnosis, some people live five or ten years, or at times even longer, although with marked disability. One of the most famous people with ALS, the physicist Stephen Hawking, lived with the disease for fifty-five years after being diagnosed at age twenty-one. No one knows why he lived so long; perhaps he had a variant of ALS. Unlike Lou Gehrig, whose muscle strength was integral to his playing baseball, Hawking, whose livelihood depended on cognition, continued his work as a physicist, although he was paralyzed.

Merit spent well over an hour talking to Tanya, discussing the path forward. There were one or two more tests to be done to rule out other conditions, but Merit was fairly certain of the diagnosis. Fortunately, there were treatments that could be prescribed. Riluzole, a drug that affects the excitatory neurotransmitter glutamate, has been used in ALS since its FDA approval in 1995. It works by decreasing the toxic effects of excess glutamate. Unfortunately, it extends survival by only a few months. Patients who have the bulbar form of ALS, like Tanya, sometimes have emotional lability, manifested by rapid mood changes, such as uncontrollable periods of laughing or crying. Nuedexta was approved in 2010 and helps these emotional lability symptoms. It may also aid some patients with speech and swallowing.

Merit then spoke to Tanya about research studies and clinical trials. Some of the trials were placebo controlled, and patients, of course, found it difficult to contemplate not knowing if they were receiving the experimental drug or not. In other trials, such as one being conducted at Mass General with a drug called ibudilast, all patients received the drug. One of the research studies involved ALS patients giving blood to better understand the disease. Tanya looked to me for guidance, and I told her it was a good idea—in fact, some of the blood might end up being tested in my own laboratory. She signed a consent form, and twelve tubes of blood were drawn.

Merit emphasized how important it was not to lose weight. ALS patients who lose weight do not do as well, perhaps as a consequence of how body metabolism affects the motor neurons. Careful measurements of breathing and swallowing would be performed to track the course of the disease and to provide help with speech and breathing. Genetic testing would be done to determine if there was a familial component to the ALS. This had become important, given that new treatment trials were being undertaken to silence abnormal genes. Most importantly, Merit explained that a new drug called edaravone had shown positive results in ALS studies in Japan and was to become the second FDA-approved disease-modifying drug for ALS in twenty years. The fact that a new drug would soon be approved for ALS offered hope, although the effect of the drug on survival was not dramatic. Finally, Merit told Tanya about stem cell therapy that was being given to ALS patients in South Korea, and that some patients traveled overseas to receive treatment.

Tanya returned to Boston three months later. She had not experienced any worsening in her disease, and her exam was unchanged. A plan was put into place. She would enter the ibudilast trial at Mass General, and we would arrange for her to be treated with edaravone. Ibudilast, given as a pill, had been used for a number of years as a treatment for asthma and stroke. It had anti-inflammatory properties and a neuroprotective effect. It was also being tested in MS with some positive effects. By being in the trial, Tanya's ALS would be carefully monitored, and she would undergo PET imaging of her brain to measure microglial cell activation. At that time, edaravone was not yet approved by the FDA; it had to be purchased and shipped from Japan. Edaravone was given by intravenous infusion in two-week cycles, and Tanya would have a port placed for ease of administration. Lazar learned to administer the edaravone, although he hated needles. Tanya had no interest in traveling to South Korea to receive stem cells. I suggested adding Biotin, a vitamin that appeared to help patients with progressive MS and that was being tested in ALS. For the time being, Tanya was at peace with her disease. A treatment plan was in place, and there was hope.

In the winter of 2017, my wife and I spent time with Tanya and Lazar in Orlando and visited the Kennedy Space Center together. The ibudilast appeared to be helping, and the edaravone would be started in the spring. However, on her next visit to Boston in the summer of 2017, it was clear that Tanya's speech was worsening. In October 2017, the ibudilast was stopped because of gastrointestinal side effects. Her speech continued to worsen, and at her June 2018 visit to Boston, she had more difficulty swallowing and had lost weight. Special milkshakes were prescribed.

Liz wanted to bring her to Colorado to live, but Tanya found it hard to breathe at the high Colorado altitude. Placement of a feeding tube was discussed, but Tanya refused until after our granddaughter's bat mitzvah in November 2019 in Denver. The family took pictures with Tanya, a beautiful Russian woman, at the synagogue. In the winter of 2019, my wife and I visited Tanya and Lazar in Florida and went to dinner. She needed help to walk, couldn't eat, and required an iPad to communicate. Breathing was more labored, but she didn't want a tracheostomy. She developed abdominal pain that we found difficult to control. Lazar was now caring for her around the clock, giving her six feedings a day through the feeding tube. When Liz flew to visit, her mother wrote on her iPad, "I want to die." Tanya passed away in August 2019, three years from the time I took the call from the neurologist in Denver while returning from my golf game.

As I sat at Tanya's funeral, I thought about how little I'd been able to help her, although I was working full steam in the lab to develop a treatment for ALS. It was the same feeling I'd had when my mother developed Alzheimer's disease and I couldn't change its course. I vowed to redouble my efforts and not be discouraged by the inevitable failures I would face in trying to find a cure for these diseases.

I thought of sitting with Tanya in a hospital exam room at Mass General, and then I thought back to 1980, when Steve Hauser and I walked the halls at Mass General and embarked on successfully treating multiple sclerosis with the chemotherapy drug cyclophosphamide. At that time, another neurology resident named Bob Brown had also finished his neurology training at Mass General, and he joined my lab as a research fellow. Bob was specifically interested in ALS. At that time, no FDA-approved drugs existed for either MS or ALS. Since then, Steve and Bob have gone on to become department chairmen and prominent neurologists. For MS patients, there are now fifteen FDA-approved drugs that shut down MS attacks. Unfortunately, for ALS patients, there are only three FDA-approved drugs, and they are not significantly effective. Bob told me when we recently sat down to talk about ALS, "Howard, I have MS envy."

THE CRIME SCENE

ALS is a disease of the motor system. The motor system consists of motor neurons that run from the top of the brain to the spinal cord. A neuron is a nerve cell that is the basic building block of the nervous system. It transmits information between one nerve cell and another by both electrical and chemical means. There are specialized neurons for every brain function, such as vision, sensation, and cognition. In ALS only the neurons related to motor function are affected. When I want to move my arm, neurons in the motor section of the brain fire and send signals down through the spinal cord, where they connect with another series of neurons, which then leave the spinal cord and stimulate my arm to move. There are motor neurons for the brain stem that stimulate muscles crucial for speech and swallowing in addition to those that stimulate muscles in the arms and legs. Until late in the disease, ALS does not affect neurons responsible for eye movement or bladder function. However, the motor neurons responsible for breathing are ultimately affected, which is the usual reason people succumb to the disease. People with trouble swallowing or speaking can have a feeding tube put in place and use a computer or other device to communicate. However, when breathing is impaired, the only way to survive is to have mechanical ventilation by a tracheostomy. Most patients do not choose to have this done.

If one looks under the microscope, the crime scene shows death of motor neurons all the way from the brain, through the brain stem, and down to the spinal cord. Because of the degeneration of the motor nerve fibers, the spinal cord is thinner, and the remaining motor neurons are shriveled. There may also be scarring, or sclerosis.

In addition to death of the motor neurons, the crime scene has a number of coconspirators. Other cells in the nervous system become affected: microglial cells, oligodendrocytes, and astrocytes, which are three of the four major cell types in the brain we discussed in the first chapter. In ALS, microglia take on the Mr. Hyde inflammatory form, which we have seen is also part of the crime scene in MS and Alzheimer's disease. The

AMYOTROPHIC LATERAL SCLEROSIS

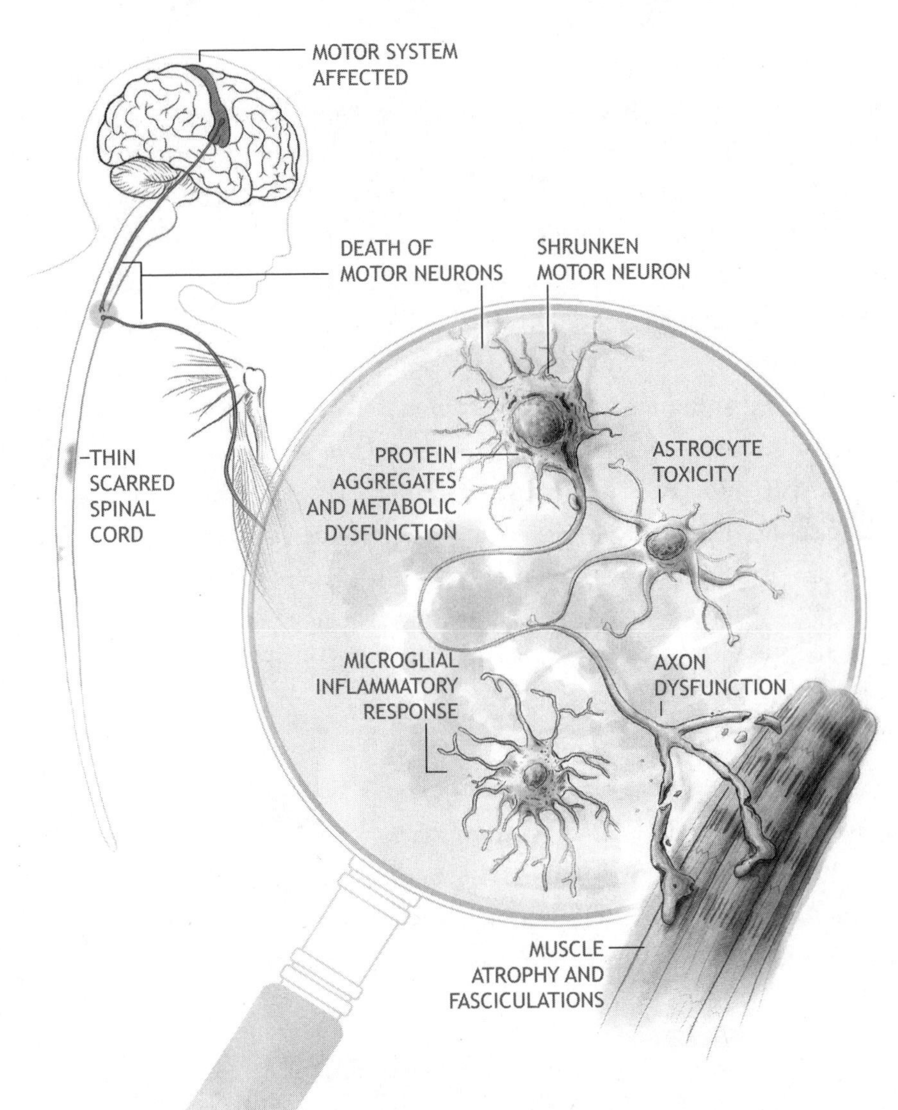

The ALS crime scene shows death of motor neurons all the way from the brain, through the brainstem, and down to the spinal cord. Microglial cells, oligodendrocytes, and astrocytes are also affected.

hippocampus, which is part of the crime scene in Alzheimer's disease and causes memory loss, is not affected in ALS. In the past two decades, however, we have recognized that, on occasion, some people with ALS who do not have pure motor disease may have problems with cognition.

As we will learn later in this chapter, these non-neuronal coconspirators also play an important role in the progression of the disease and may be targets for therapy. In addition to death and thinning of the motor neurons, many of the neurons have clumps of a protein called TDP-43 inside them, which can sometimes be seen under the microscope. Protein aggregation is also prominently seen in Alzheimer's disease and Parkinson's disease.

MS has strong environmental factors, but the environment does not play as much a role in ALS, although there is an increased risk of ALS in people with military service, those who smoke, have been exposed to heavy metals or pesticides, have experienced head trauma, and perhaps even those who have engaged in vigorous exercise. There is no evidence of either an infectious agent or the ability to transmit ALS from one person to another, although early studies considered the possibility that ALS was somehow related to poliovirus or other viruses.

As we will see, the immune system may also be a coconspirator. However, simply suppressing the immune system doesn't help ALS. When Steve Hauser and I found the chemotherapy drug cyclophosphamide helped MS, Bob Brown suggested we try it in ALS, but it didn't work. Others have tried strong suppression of the immune system in ALS, including something called total lymphoid irradiation, but, again, to no positive effect.

GENETICS AND ALS

A major breakthrough in our understanding of ALS came from genetics. Bob Brown made the discovery. When Bob was in my laboratory, he considered many potential avenues by which to investigate and understand the biology of the motor neuron. What were its features? What was on its surface? How did it die? Perhaps polio, which is also a paralytic disease of the motor neuron, could give a clue—the poliovirus binds to the surface of the motor neuron. Coincidentally, we were actually working in a building named for John Enders, who won the Nobel Prize in Medicine for the discovery of the poliovirus. In another connection, my house in Brookline, Massachusetts, is built on land that was formerly the gardens of John Enders's house. However, we could find no link between poliovirus and ALS.

Understanding the biology of ALS turned out to be an incredibly difficult challenge. After Bob left the lab, he reasoned he could bypass the need to understand the cell's biology by finding out what gene was causing the disease. There are natural variations throughout the genome, and it was possible to track DNA variation in families with ALS. To make the method work, nothing about the disease needed to be known except that it was inherited. So, Bob did the opposite of hypothesis-based disease discovery. Instead, he said, "Give me a family pedigree and gene variance, and to hell with knowing the exact cause. I'll find a gene linked to the disease, and understanding what that gene does will lead to a cure."

In October 1992, his team made a major breakthrough. He was at a meeting in Japan when he received an urgent message to call his lab in Boston. When Bob returned the call, he learned that a laboratory gel they had set up to examine genetic material in ALS was positive. The horizontal bands on the gel showed that normal DNA and DNA from people with familial ALS were dramatically different. They had found a gene linked to ALS.

Bob knew at that time the result was both genuine and a potential game changer. It was a eureka moment for him. He stayed up all night in Japan, figuring out what to do next. As he confirmed the finding and began to write the paper, he called the editors of *Nature* to tell them what he had found. They were almost as excited as he was. They had been hoping someone could find a gene related to ALS. Bob submitted his paper on February 11, 1993. It was accepted for publication five days later and published on March 4, twenty-two days later. The discovery of DNA by Watson and Crick was published twenty-four days after its submission to *Nature* on April 2, 1953.

The title of the paper was "Mutations in Cu/Zn Superoxide Dismutase Gene Are Associated with Familial Amyotrophic Lateral Sclerosis." The Cu/Zn superoxide dismutase gene was abbreviated *SOD*. There were thirty-three authors on the paper, and Bob was the last, or senior, author. The abstract was simple and straightforward, describing ALS as a uniformly fatal motor neuron disease, its cause unknown, with 10 percent of people with ALS having a strong genetic component. They were autosomal dominant, meaning only one mutant gene was needed to cause ALS. Bob reported a tight genetic link between familial ALS and the gene encoding SOD. He identified *SOD* as a candidate gene in familial ALS and eleven different mutations of *SOD* in thirteen different ALS families. Finally, there was a smoking gun at the crime scene. The *SOD1* gene coded for a major antioxidant protein that detoxifies superoxide radicals, so in the case of the mutation, there was a defect in being able to clear a toxic substance from the cell.

Keep in mind that the abnormal accumulation of the SOD protein in motor neurons also occurs in other cells of the body, but it doesn't damage those cells. The reason for this selectivity may be that the motor neuron is one of the largest cells in the body. It has to support the survival of an enormously long nerve process, and it has a limited capacity to regenerate, which means the motor neurons you're born with will be with you your entire life. The accumulation of a toxic SOD protein could be lethal.

Talking to me over dinner at a delicatessen in Brookline, Massachusetts, Bob said, "If the room we are sitting in was a motor neuron and it had to communicate to the muscle through its nerve fiber, the muscle would be in Manhattan." Thus, the motor neuron may be especially vulnerable when something goes wrong. Clearly, the heart of the crime scene is the motor neuron, but as we will learn, there are coconspirators.

Just after Bob made his discovery of SOD in familial ALS, Merit Cudkowicz was a neurology resident at Massachusetts General Hospital. In 1994, she and Bob immediately began working together on a clinical trial to take advantage of the breakthrough. Merit thought that trials of experimental treatments for people with ALS were finally feasible. With Bob and a few other investigators, they formed the Northeast ALS Consortium, a trial network to bring treatments to people with ALS. Now with over 120 sites, the consortium is pioneering innovative trials for people with ALS.

At the time they started the consortium, Bob had identified a family in which four of five members carried the SOD mutation. Three of them became sick with ALS almost simultaneously. If the SOD protein was defective in these ALS patients, Bob and Merit reasoned that giving them back a normal SOD protein would be of benefit. She and Bob administered a healthy SOD protein, taken from a cow, to the three family members and about a dozen others who had ALS and also carried the SOD genetic mutation. The protein was injected into the spinal canal. They followed the patients carefully and watched hopefully to see if the disease would be arrested. Sadly, the treatment didn't work. Over the next two to three years, all the patients died. Nonetheless, both Bob and Merit have pursued treating patients who have the genetic SOD mutation, only now they do it by trying to *decrease* its toxic function by genetic means rather than by giving more SOD. Twenty years from Bob's discovery of the SOD1 mutation, great progress has been made.

Merit cares for the next generation of the family who received the first SOD experimental treatment. She has devoted her career to caring for ALS patients and designing clinical trials. Merit told me she learned everything she needed to know about ALS from that first trial she conducted with Bob Brown. She also learned about caring for ALS patients as she spent a lot of time with the family. Every doctor I speak to talks about the pain of taking care of ALS patients and watching them succumb to the disease. Research provides hope not only for the patients but for the doctors as well. Many of them couldn't continue to treat without that hope.

THE ORIGINS OF ALS

The discovery of the SOD mutation in 1993 explained some cases of familial ALS and finally provided an animal model to study the disease. But it did not fully explain the origins of ALS. It took more than two decades for a second major genetic discovery to shed new light on ALS and to identify the most common genetic abnormality in ALS, called *C9orf72*.

The discovery of *C9orf72* stemmed from the relationship between ALS and a type of dementia called frontal temporal dementia, or FTD. FTD causes nerve cells in the frontal and temporal lobes of the brain to shrink, affecting behavior, personality, and language. It is known that ALS sometimes affects not only the motor neurons but also frontal and temporal functions, including cognition and behavior, occurring in up to 50 percent of patients. Similarly, as many as half of FTD patients develop clinical symptoms of motor neuron disease. Scientists hypothesized that ALS and FTD might be part of the same disease spectrum. In support of this theory, there is an abnormal protein called TDP-43, which is found not only in nerve cells of the majority of ALS patients but also in common subtypes of FTD.

A number of studies suggested that researchers looking for a genetic connection between FTD and ALS should focus on chromosome 9. In 2011, two independent groups provided evidence that the defect connecting FTD and ALS was, in fact, on chromosome 9. One group studied families in Vancouver, San Francisco, and the Mayo Clinic in Rochester, Minnesota, while the other group studied families in the United Kingdom and the Netherlands.

By examining chromosome 9, they found an area where the same genetic material was repeated over and over, like a record that gets stuck and plays the same line of a song over and over. This is significant because four bases make up the genetic code of our DNA: A, C, G, and T. In our genome, there is a repeat of six bases, called a hexanucleotide repeat, comprising the following base sequence: GGGGCC. In healthy individuals, the sequence GGGGCC is repeated two to twenty-three times. What investigators found is that in individuals with ALS or FTD, these repeats occurred by the hundreds or even thousands. This genetic abnormality occurred on chromosome 9 at open reading frame 72. Thus, the gene is called *C9orf72*.

The major question then was, How did the alteration in the *C9orf72* gene contribute to ALS? There were two possibilities. The first was that the *C9orf72* gene was carrying out an important function in the cell, and the defect in the gene disrupted a normal pathway needed for cell function. This is called loss of function, and the function that is lost causes the disease. For example, if you lose your eyes, you lose visual function and the ability to see. The second possibility was that the alteration in the *C9orf72* gene contributed to ALS because of a gain of function. In other words, the gene introduced a new function to the cell, a function that was toxic to the motor neuron and thus promoted the disease.

To test whether *C9orf72* caused disease because it caused a lack of function, researchers created mice that lacked the gene. However, mice that had no *C9orf72* gene in any cell in the body developed and aged normally, with no motor neuron disease. They did, however, develop changes in their spleens and lymph nodes, and had abnormalities in white blood cells and brain microglia, suggesting that the role of the *C9orf72* gene in ALS could relate to the non-neuronal cells. However, given all the evidence, it appeared more likely that a gain of toxic function was the main mechanism by which the abnormality of the *C9orf72* gene led to ALS. In other words, it wasn't because something was missing or because the gene wasn't functioning appropriately. It was because the hundreds of thousands of repeats of GGGGCC made the *C9orf72* gene more toxic.

In support of this concept, when investigators delivered the expanded GGGGCC repeats into the brains of normal mice, there was degeneration in nervous system cells, analogous to what was seen in ALS. There were several possibilities for why the expanded GGGGCC repeats were toxic, all related to the basic metabolism in the cell, including the mutant gene disrupting the movement of proteins within the cell and transportation of crucial molecules from one part of the cell to another to maintain normal cell function.

Regardless of the precise manner in which the *C9orf72* gene was toxic to neurons, the mutant gene provided a new target for therapeutic invention and a tool to screen for those at risk for developing ALS. As with *SOD*, mouse models that express the human *C9orf72* gene and contain about 450 GGGGCC repeats are being developed, and they repeat the pathological and other features of what is called C9ALS/FTD. Recognition of the *C9orf72* gene has also allowed prognosis and characterization of ALS patients.

It is clear that ALS may represent different subtypes. As in breast cancer, where people with different genetic mutations respond to different therapies, the same may ultimately be true in ALS. A study of almost five thousand ALS patients is underway to determine if the *C9orf72* gene is linked to sub-populations of the disease or to sex differences.

Because the abnormalities in the *C9orf72* gene have been associated with familial frontal temporal dementia in which there are cognitive problems, the presence of the *C9orf72* repeat expansion was investigated in a large number of Alzheimer's patients. However, it turned out that very few of the Alzheimer's patients had the *C9orf72* repeat, suggesting it is not a feature of Alzheimer's disease.

One possible approach for translating the abnormalities in the *C9orf72* gene into therapy for ALS is to dampen the abnormal gene using an ASO, or antisense oligonucleotide, as a form of gene therapy. ASOs have been successfully used to treat a lethal muscle disease of children called spinal muscular atrophy, which we discuss in detail later. Briefly, an oligonucleotide is a short string of nucleic acids. *Antisense* means you are providing nucleic acids that complement the segment you are interested in. For example, if you had GGGGCC, then the antisense string would be CCCCGG because cytosine (C) matches with guanine (G) and vice versa. Some of these pathological abnormalities in ALS could potentially be reversed with an ASO that degrades and decreases the RNA encoding the C9orf72 protein. In animal models, investigators have been successful in treating disease caused by the abnormal *C9orf72* gene using ASOs that target excess GGGGCC repeats, and a clinical trial is underway to test this approach in ALS patients.

We know there are both familial and nonfamilial, or sporadic, forms of ALS, but genetics has led the way to our understanding of the origins of the disease ever since Bob Brown's lab found the *SOD*-ALS link. There are now more than fifty genes that have been shown to be related to ALS. The *C9orf72* gene is found in 25 percent of familial ALS patients and 10 percent of sporadic ALS patients, and 1 to 3 percent of sporadic cases have mutations in the *SOD1* gene. Because familial and sporadic ALS look so similar clinically, some wonder whether genetic variants are responsible for all ALS and we just have not identified all the genes responsible.

What do these genetic mutations do, and how do they affect the cell, leading to such a horrible disease? Although the answer is not known definitively, it is generally felt that vital cellular processes are disrupted in three areas.

First, in the cells of ALS patients, there is an accumulation of aggregated proteins and defects in pathways for breaking down aggregated proteins. Mutant versions of *SOD1* often form intracellular aggregates that clog up the cell. A number of other genetic factors cause protein abnormalities. In this way, ALS is similar to other degenerative diseases such as Alzheimer's and Parkinson's disease, in which there is an abnormal accumulation of proteins in the cell.

The second cellular process implicated in ALS relates to RNA. RNA is the messenger that carries the genetic information from DNA to make proteins. In ALS, it appears that there are several abnormal genes that disrupt RNA metabolism. The disrupted RNA metabolism leads to abnormal proteins. The first such protein to be discovered is called TDP-43. Mutations in TDP-43 are present in a number of neurodegenerative diseases. The most common mutated gene in ALS, *C9orf72*, also plays a role. The hundreds or thousands of repeats found in the abnormal *C9orf72* gene can form RNA that generates proteins that are toxic to the cell.

The third cellular process that disrupts the motor neuron relates to the scaffolding or cytoskeleton that holds the cell together. Three ALS genes have been identified that disrupt proteins that are important in the normal scaffolding of the cell. These genes have also been identified in some sporadic cases of ALS.

IN SEARCH OF ADDITIONAL CULPRITS

Although most ALS patients do not have the abnormal SOD mutation that Bob Brown discovered, the mouse model of ALS with the SOD mutation has become a mainstay for investigating the disease. Additional mouse models have been generated using the other genes that have been identified. After these mouse models were created and clearly mimicked damage to the motor neuron that occurs in ALS, a major surprise in our study of ALS occurred with the discovery that it isn't just an abnormality of the motor neuron that drives the disease: other cells are involved. Pioneering studies by Don Cleveland at the University of California San Diego and others showed that cells which are *not* motor neuron cells are important for driving the disease. This discovery opened up potential new avenues of therapy.

In a groundbreaking study published in 2006, Cleveland took the SOD mouse and, through sophisticated genetic manipulation, was able to create a mouse in which the abnormal *SOD* gene was present in the motor neurons but not in the microglial cells surrounding motor neurons. It was presumed that the toxic effects of SOD were due to the presence of abnormal SOD in the motor neurons, so removing the SOD from the microglial cells should have no effect on the disease. It was a little bit like asking, What's important for vision, your eyes or your mouth? Cover up your eyes and you can't see; cover up your mouth and you can still see. In the study, they were investigating if taking away the mutant SOD from the microglial cells would affect the motor neuron and disease progression.

The surprising answer was yes. Cleveland found that although decreasing SOD in the motor neurons delayed the *onset* of the disease, removing the SOD from the microglia also had an effect. Decreasing SOD in microglia did not affect disease onset but dramatically slowed disease *progression*. Thus, the abnormal SOD was doing something that impinged on the function of the microglia. In other words, dysfunction of other cells next to the motor neuron enhanced motor neuron damage, and microglial cells become activated at or before disease onset in the SOD mouse. This is an important discovery, because these experiments raised the possibility that targeting the microglial cell might help patients with ALS. Although the onset of ALS resulted from damage to motor neurons, Cleveland's findings showed therapy could be successful by targeting a cell that was not a motor neuron. A Mr. Hyde–type microglial cell was a coconspirator and was contributing to the progression of ALS.

Further proof of Cleveland's finding came from Stanley Appel's laboratory at the Houston Methodist Hospital. Appel found that when normal microglia were transplanted into an ALS mouse, the normal microglia slowed motor neuron loss and animals lived longer. Understanding the important role of microglia in maintaining the health of the motor neuron also came from studies in our laboratory in which we removed microglia from the nervous system of normal mice and found that the animals developed an ALS-like disease without having an abnormality in the *SOD* gene.

Because microglia play a role in maintaining the health of motor neurons, these studies raise the possibility that treatment targeting microglia may in fact help ALS. They also raise the possibility that if microglial cells are involved in the genetic, or familial, ALS, they could also be involved in sporadic ALS, which would provide the opportunity for a treatment that would apply to all ALS patients, not only those with a specific gene abnormality.

Microglia are not the only non-neuronal cells in the brain that contribute to ALS. Another major non-neuronal cell in the brain is the astrocyte, and like microglial cells the astrocyte can also be toxic to motor neurons. In addition to their experiments on microglia, Don Cleveland and his group did an identical experiment in which they showed that if they decreased

SOD in astrocytes, as with microglia, they also delayed progression, though disease onset was not affected.

One of the major questions in ALS is the degree to which sporadic ALS is the same as familial ALS in which people have the abnormal *SOD* gene. Studies have shown in fact that astrocytes from both familial and sporadic ALS are toxic to motor neurons. This raises the possibility that therapies targeting astrocytes or microglia may benefit all types of ALS. Indeed, scientists have reported that if both astrocytes and microglia are targeted, there is an additive protective effect.

The other major cell type in the brain that is not a neuron that may play a role in ALS is the oligodendrocyte. The oligodendrocyte is the cell that wraps around the nerve fiber or axon in the brain and makes myelin. It plays a very important role in MS, in which the myelin sheath is attacked by the immune system. Investigators, led by Jeffrey Rothstein at Johns Hopkins, have found that the oligodendrocyte does more than provide myelin to the axon. It provides energy that is important for axon survival and may have an important role in motor neuron survival in ALS. A structure in oligodendrocytes called MCT1 releases lactate, providing energy to motor neurons. Lactate is a toxic substance that builds up after exercise—the muscle stiffness people feel is related to lactic acid. It appears that an important transporter of lactate may be dysfunctional in ALS, because reduction of MCT1 in oligodendrocytes induces motor neuron death.

The discovery of SOD has been a transformative event in our attempts to understand and treat ALS. The SOD mouse model is one of the most commonly tested models of the disease because it looks like ALS both pathologically and clinically. In people, trials to genetically target SOD are underway. Although for most sporadic cases there is no known genetic defect, sporadic cases look clinically and pathologically like the familial cases, although those with familial or genetic forms of ALS have a more virulent disease.

We now believe that the SOD mutation causes motor neuron degeneration because it disrupts the machinery of the motor neuron, which, because of its long process and energy requirements, is especially vulnerable to damage. We also know that the crime scene in ALS is not limited to a shrunken and damaged motor neuron but also involves the surrounding microglia, astrocytes, and oligodendrocytes, which act as coconspirators and contribute to motor neuron damage. After Bob Brown discovered the *SOD* gene, and Bob and Merit Cudkowicz treated patients with normal bovine SOD, we now know that the SOD is toxic and the mutant SOD damages the motor neurons. Almost thirty years after his discovery of the *SOD* gene, Bob Brown and others have begun to treat ALS using a gene therapy approach to decrease toxic SOD in familial cases of ALS. As we will discuss, initial trials are promising.

IMMUNOLOGY AND ALS

Every scientist brings their own area of expertise to study disease. Bob Brown's discovery of SOD was the result of a genetic approach to understand ALS. A chemist would look at chemical compositions of the diseased brain, and a radiologist would focus on different ways to image the brain. Of course, every scientist hopes that their approach will shed light on a key biologic finding that drives the disease.

With my expertise in immunology gained from my study of MS, I have investigated ALS by applying immunologic approaches to understand and treat the disease. However, global immunosuppression of the immune system with cyclophosphamide did not help ALS, as it did in MS, and few, if any, immune genes are linked to ALS. Nonetheless, there is ample evidence that the immune system could play a role in ALS. Inflammatory changes are seen in the spinal cord of ALS patients, and it isn't only motor neuron cells but other cells such as microglia that contribute to the degeneration of the motor neuron.

When Oleg Butovsky joined my laboratory, we developed new tools to understand the difference between microglial cells in the brain and white blood cells called monocytes that enter the brain from the bloodstream.

Monocytes are white blood cells with scavenger function. We took techniques developed for the study of monocytes in MS and applied them to ALS. We found there were changes in the monocytes in the spleens of mice with the SOD type of ALS. Furthermore, as the disease progressed and motor neurons were damaged, the monocytes infiltrated the spinal cord. More interestingly, the spinal cord microglia aided in the recruitment of monocytes, which enhanced the damage. The big question, of course, was whether targeting the monocytes in the spleens of SOD mice would impact disease progression. Would it damage the motor neurons in the spinal cord?

To answer the question, we developed an antibody that targeted monocytes and injected it into the bloodstream of mice. We found reduced movement of monocytes from the spleen into the spinal cord and less damage to motor neurons, and the animals lived longer. Of course, as we all know, what is found in mice is not always replicable in humans. Animal models are crucial for our research, but a link must be found between how a treatment functions in a mouse and how that will play out in people. To test this in humans, we isolated monocytes from the blood of patients with ALS; to our excitement, we found changes in monocytes in their blood that were similar to those we observed in the spleens of mice. Just like the SOD mice in our experiments, inflammatory genes in blood monocytes of ALS patients were elevated. Interestingly, some of the inflammatory genes in the monocytes of ALS patients were also observed in our MS patients, indicating that inflammatory processes played a role in both diseases. MS is primarily considered an inflammatory and autoimmune disease. Our results suggested that there was a secondary inflammatory or even autoimmune component in ALS as well.

More importantly, when we looked at the monocytes of patients with sporadic versus familial genetically driven ALS, we found changes in both. This was good news, because it raised the possibility that treating inflammation could help all people with ALS, independent of the type of disease.

In 1953 Watson and Crick discovered the double helix structure of DNA and how genetic information in the cell is processed. DNA is transcribed to RNA, which is then translated to a protein. Scientists have since learned that the genetic control of biologic processes is much more complex. We now know that there are small RNAs, called microRNAs, that do not encode proteins but serve to regulate gene function. In 1993 the first microRNAs were reported in worms by Victor Ambros and Gary Ruvkun. They received the Lasker Prize for their work in 2008 when it became clear that they had discovered a new genetic form that played a key role in cell biology and offered a new avenue to treat disease. MicroRNAs provided a whole new lens with which to look at disease. When Bob Brown and I were studying ALS in the '80s, microRNAs hadn't been discovered. It makes one wonder whether there are aspects of biology yet to be discovered that will be required to solve a disease such as ALS.

Oleg Butovsky and I found abnormalities in a microRNA called miR-155 in both the monocytes and microglia of SOD mice with ALS. miR-155 is microRNA that promotes inflammation that we had also found was elevated in MS.

The next question we faced was whether the miR-155 was actually driving the disease in ALS mice. One of the most powerful tools we have to study disease mechanisms in mice is the ability to specifically delete, or knock out, a single gene using genetic engineering. We did so by creating a mouse that had every gene except *miR-155*. Knocking out miR-155 in a normal mouse did not affect the survival of the mouse, thus the path was open to determine the effect of knocking out the *miR-155* gene in the SOD mouse. To delete the *miR-155* gene in the SOD mouse, we mated a miR-155 knockout mouse to an SOD mouse, which in effect created an SOD mouse that didn't have the *miR-155* gene. To our great excitement, the SOD mouse without the *miR-155* gene lived much longer. I remember Oleg calling me from the mouse colony and saying, "They're still alive, Howard, and they are running normally." Furthermore, knocking out miR-155 changed and reversed the abnormalities we found in the microglia and monocytes of the mice.

It is not possible to treat people by knocking out a major gene, so we decided to determine if we could use a synthetic compound to selectively decrease miR-155 in ALS mice. We were able to do so. We found that a synthetic compound called anti-miR-155, which specifically blocks miR-155, delayed disease onset and extended survival in SOD1 mice. Similar results were obtained by Tim Miller at the University of Washington. It is always gratifying when another laboratory independently confirms one's findings.

Targeting miR-155 had the potential to serve as a treatment for ALS, but there was one more hurdle. We found that miR-155 was elevated in mice with ALS, but we didn't know if it was elevated in patients with ALS. If not, there wasn't much logic to using anti-miR-155 to treat people. Ultimately, we found that miR-155 was elevated both in blood monocytes of ALS subjects and in their spinal cords. Additionally, it was elevated both in sporadic and familial forms of the disease, making anti-miR-155 a broadly applicable therapy. We envision injecting the anti-miR-155 into the spinal canal and are in the early planning stages to test anti-miR-155 in patients.

The immune system has a mechanism to suppress inflammation. As we discussed in the chapter on MS, there are regulatory T cells that dampen inflammatory T cells. Stanley Appel at Baylor University has argued that regulatory T cells are defective in ALS, just as they are in MS. Boosting regulatory T cells would be of benefit in ALS. An initial trial Appel performed with Merit Cudkowicz and the Mass General team to boost regulatory T cells suggested some potential benefit, and he has recently carried out a trial in which he grew regulatory T cells in the test tube and then injected them into ALS patients. Those patients who received the regulatory T cells appeared to have slowing of their disease. The regulatory T cells presumably acted by dampening inflammation in the brain and spinal cord. Based on our studies in MS, we also are investigating ways in which to induce regulatory T cells to treat ALS patients. As discussed in the MS chapter, the nasal anti-CD3 monoclonal antibody we are planning to test in progressive MS induces regulatory T cells. We have also found it prolongs survival in ALS mice, and we plan to test it in ALS. Another shot on goal.

Nonetheless, some trials of anti-inflammatory agents have not been successful in ALS although they showed positive results in animal models.

The reason is not known but probably relates to the stage of disease when the drugs were given, the subcategories of patients, and the complex nature by which the immune system is regulated. We know, for example, that there can be opposite effects when immune treatments are given. For example, in MS a drug that targeted the chemical TNF worked well in the animal model of MS but unexpectedly made MS worse when it was tried in people with the disease.

To modulate the immune system or any target in ALS, there needs to be a measure that can be used to demonstrate that the treatment actually works. Indeed, there are studies that show that changes in the immune system that can be measured in the blood occur in ALS and correlate with rapid ALS progression. Additionally, if microglial cells in the brain are involved, as we think they are, it is crucial to show they are inflamed in patients with the disease and that they can be quieted with treatment.

It is possible to measure microglial activation by PET imaging of the brain, a technique used in diseases such as MS and Alzheimer's disease. In studies by Nazem Atassi at Mass General, microglial activation as measured by PET imaging is also seen in ALS. In a clinical trial of ibudilast, PET imaging of microglial inflammation is one of the measures of success. Although the ultimate proof of a treatment is its ability to reduce symptoms or slow progression, the ability of a radiologic or blood or spinal fluid biomarker to show how a treatment affects the disease processes is critically important. It is like a chest X-ray for tuberculosis or measuring blood pressure for a drug that then translates into fewer heart attacks. Treating inflammation in ALS is likely to be part of combination therapy that ultimately controls the disease.

MEASURING DISABILITY IN ALS

As with MS and the other diseases discussed in this book, investigators are attempting to develop an objective clinical functional scale by which to measure the degree of disability a patient has and to serve as an outcome measure for trials. For MS, we use the EDSS, or Expanded Disability Status Scale; in Alzheimer's disease, it is the ADAS-Cog, or the Alzheimer's Disease Assessment Scale–Cognitive Subscale. One of the most commonly used functional scores in ALS is the ALSFRS-R, or the ALS Functional Rating Scale-Revised. The ALSFRS-R consists of twelve items across four measures of body function, including bulbar (swallowing), fine motor, gross motor, and breathing. Each item is rated from zero to four, with zero being loss of function and four being normal function. A perfect score is 48. Patients with ALS typically decline by approximately one point per month on average, so the ALSFRS-R captures clinically relevant features of disease progression and correlates them with survival. This scale has been used widely in Phase 1 and Phase 2 trials. However, like the EDSS for MS, there is subjectivity in scoring patients, and each number on the scale does not carry equal weight in terms of disability. Furthermore, measures such as muscle strength and tracheostomy-free survival are not particularly helpful or descriptive in trials of patients who are in the early stages of the disease. Objectively measured biomarkers in blood or spinal fluid that directly link to the disease process are needed.

One such measure we can use for this purpose is neurofilaments, which are breakdown products of the nerve fibers in the brain. They've now been shown to correlate with damage to the brain in patients with MS, Alzheimer's disease, and many other brain diseases. Measuring neurofilaments used to require spinal fluid, but they can now be measured in blood. Investigators have found that spinal fluid and serum neurofilament levels can distinguish early symptom onset ALS from other neurological diseases. This could enhance the diagnostic accuracy in assessing ALS.

Other markers in the blood that are abnormal in ALS include plasma creatinine, which may find use as a prognostic biomarker in ALS. Creatinine is a breakdown product of muscle and is related to muscle mass. Several studies have shown that people with ALS have lower plasma creatinine levels. A study of more than a thousand patients showed plasma creatinine could be a predictor of mortality in ALS. In the urine, neurotrophin receptor p75 may also link to disease progression, as one study showed there were higher levels of urinary p75 in ALS patients than in healthy subjects and the levels rose as the disease worsened.

Inflammation plays a role in ALS, and C-reactive protein is a classic marker of inflammation. It has prognostic value in a number of diseases including tumors, cardiovascular disease, and rheumatoid diseases. It also appears that C-reactive protein blood levels correlate with degree of disability in ALS.

A study of proteins in the spinal fluid of ALS patients showed that certain proteins called chitinases were elevated and the elevation was linked to the activation of white blood cells called monocytes. As discussed, in the mouse model, monocytes are related to damage to the motor neuron. Finding elevated chitinases in the spinal fluid supports attempts to treat inflammation in ALS.

Finally, in a large study carried out in Sweden, more than half a million people had their blood analyzed for biomarkers, including carbohydrates, lipids, and apolipoprotein metabolites to determine if any were associated with ALS risk. Some associations between ALS and cholesterol and apolipoproteins were identified, and this epidemiologic data, collected over twenty years, could help identify factors associated with ALS risk that can be measured prior to the diagnosis of ALS. Blood tests and spinal fluid examination will be needed to guide treatment and develop personalized approaches for the treatment of ALS.

ALS AND GENES

The identification and investigation of the genes linked to ALS have major implications for ultimately understanding the disease, for clinical counseling, and for treatment. Zou Zhangyu and his group at Fujian Medical University Union Hospital in Fujian, China, reviewed over six thousand articles dealing with the frequencies of the major four genes in ALS: *SOD1*, *C9orf72*, *TARDBP*, and *FUS*. We have already discussed *SOD1* and

C9orf72. *TARDBP* is a gene that codes for a protein that binds to DNA and is involved in transcription, the first step in protein production. *FUS* is a gene that provides instructions for a protein that binds to DNA and RNA and is also involved in protein production.

European and Asian ALS cohorts show divergent genetic backgrounds. In European patients, both sporadic and familial, the gene *C9orf72* is the most common, whereas the *SOD1* gene is most common in Asian patients. This shows that the genetic origin of ALS is not uniform across the globe. Also, there is a strong indication for the occurrence of two gene mutations in the major four genes. Twin studies have shown that in identical twins, it is very uncommon for the unaffected twin to develop the disease, unlike in MS, suggesting environmental factors play a role. ALS is discordant in 90 percent of identical twins.

An important question is what percentage of isolated or sporadic forms of ALS have identifiable genetic factors that are likely responsible for the disease. Summer Gibson and his team from the University of Utah School of Medicine studied eighty-seven patients of European ancestry diagnosed with sporadic ALS and focused on thirty-one genes that had been linked to ALS, such as *SOD*, as well as two of the sequences in *C9orf72* and *ATXN2*. The *ATXN2* gene is linked to spinocerebellar ataxia and increases the risk of ALS. It turned out that only a fifth of the sporadic group had an identifiable probable genetic cause among the thirty-three study genes. This points to the complex nature of genetic contribution to ALS. There could be multiple modes of inheritance in which a gene or a number of genes can have cumulative effects or could be expressed under certain conditions. Of the sporadic ALS patients with an identifiable genetic variant, the most common was *C9orf72*, with 5.7 percent, while *SOD1* was 2.3 percent.

Given that trials targeting *SOD* and *C9orf72* are underway, it becomes crucial to understand the degree to which these genes may be involved but are only incompletely expressed in patients. Regarding *C9orf72*, healthy individuals have ten or fewer GGGGCC repeats, whereas those with ALS or frontotemporal dementia or FTD have between thirty and thousands of GGGGCC repeats. However, an intermediate repeat number of twenty to thirty has been found in both patients and controls, suggesting other factors may play a role, which makes the measurement of repeats less valuable in predicting early onset of disease.

Alice Vajda's group at Trinity College Dublin surveyed clinicians from twenty-one different countries to find out how they defined familial ALS and what their attitudes were toward genetic testing. The clinicians knew familial ALS is limited by incomplete expression of disease-causing genes and small family size, which, in turn, can result in patients with familial disease presenting as sporadic ALS. It has also become clear that the majority of future genetic advances will involve identifying many rare, novel, or low-risk genetic variants. Because of the absence of a formal evidence-based definition of familial ALS, it is hard for clinicians to advise patients. Thus, in their survey, almost three-quarters of doctors said there's no consensus for a definition of familial ALS. Most doctors offer genetic testing for the four main genes: *SOD1*, *C9orf72*, *TARDBP*, and *FUS*.

A consensus statement in 2015 issued by ALS investigators said hereditary or familial ALS should be considered if at least one first degree or second degree relative has ALS and/or FTD. This means not only parents and children but also grandparents, aunts, uncles, nephews, nieces, and half siblings. Another factor of genetic testing is the cost, which is approximately six thousand dollars. In response to the study, almost half of physicians surveyed said they would not seek genetic testing if they had personally received a diagnosis of ALS. This is analogous to those who have a family history of Huntington's disease, where there is dominant inheritance, and decide not to be tested because there's no treatment. In ALS, this perspective has begun to change with trials underway for genetic forms of ALS related to *SOD* and *C9orf72*. Furthermore, there can be different disease severity, even with the same abnormal gene. Thus, the *SOD1* gene has over 160 reported mutations, resulting in ALS that ranges from rapidly progressive to a mild, slowly progressive disease.

IN SEARCH OF TREATMENTS FOR ALS

The study of human disease involves investigating the mechanisms implicated in the disease and then determining how to interfere in these mechanisms to find an effective treatment. New technology allows us to understand biology better. We peel the onion to comprehend the layers beneath layers that we hadn't seen before. Sadly, despite advances that have been made in peeling the onion to recognize the mechanisms driving ALS, we have not made great progress in finding treatments that significantly alter the disease course.

Modern ALS trials began in the 1980s and initially investigated whether treatment of a presumed polio infection could be effective because, at that time, latent polio was suspected as a main cause of ALS. We now know this is not the case, although the poliovirus affects motor neurons, and there can be a post-polio-type syndrome that resembles ALS in some ways. Bob Brown began his work on ALS in my lab studying the poliovirus receptor, but then abandoned that line of inquiry.

Since the 1980s, nearly fifty randomized clinical trials for disease-modifying therapy have been undertaken for ALS, yet there are only three FDA-approved drugs. There haven't been more successful trials for a number of reasons. First, we may have the wrong target: the underlying hypothesis of what's driving the disease could be wrong, and thus we are not targeting the correct process that leads to motor neuron dysfunction. Second, because of difficulties in clinical trial design, we do not have measures to tell us that we have in fact hit our target. Third, some believe that testing the SOD mouse is not representative of human ALS. Finally, it may be that that treatment needs to be started at very early stages, even presymptomatic, because once someone has the disease, it's hard to stop the process—the horse is already out of the barn. To shed light on the process, we will explore the numerous trials that have been undertaken and the disease mechanisms they have targeted. What we have learned will help us one day find an effective treatment for ALS.

The first drug to be approved for ALS was riluzole. Riluzole was studied by a group at the Salpêtrière hospital in Paris. Interestingly, this is the same hospital where Charcot first described ALS. A controlled trial of riluzole in ALS was published in 1994 in the *New England Journal of Medicine*. Testing riluzole came from the hypothesis that glutamate, the primary excitatory neurotransmitter in the central nervous system, accumulates to toxic concentrations in ALS and causes neurons to die. A number of drugs are known to modulate the glutamate system, and riluzole was one of them. Following a Phase 1 trial in healthy human volunteers, single doses of riluzole of up to 200 mg were found to be well tolerated and safe.

This led the way for the randomized placebo control trial. The primary outcome measure was survival. The investigators found that after twelve months, 74 percent in the riluzole group were alive, whereas only 58 percent in the placebo group were still alive. This had a P value of 0.014. Remember, it is convention that a P value of .05 or less is considered biologically significant. Although in some sense it is an arbitrary number, it means that the results would have happened by chance only one of twenty times at most. The lower the P value, the less likely something will happen by chance. A P value of .01 means it will happen by chance 1 in 100 times, a P value of .001 is a 1 in 1,000 chance. The extended survival effect was seen primarily in patients with bulbar onset disease, which affects speaking and swallowing, as opposed to limb onset disease, which begins with arm or leg weakness. The therapeutic effect of riluzole was strongest in the first twelve months of treatment, then decreased during months twelve to twenty-one. Riluzole was approved by the FDA in 1995.

A follow-up study was done by the same group and published in 1996, in which 959 patients with ALS of less than five years' duration received either placebo, 50 mg, 100 mg, or 200 mg of riluzole daily. The investigators were able to repeat their findings and found that the 100 mg dose was most effective. This is the dose that is used today. They again found that efficacy decreased over time. However, they did not find a difference between bulbar and limb onset ALS, as positive effects were seen in both types of disease. Today riluzole is given to all ALS patients, independent of the type of disease or their age. Even if the effect of a drug is minimal, any positive effects give important clues to disease mechanism, as they demonstrate that targeting a specific biologic pathway can affect the disease. It is believed that riluzole acts on glutamate signaling. Glutamate can cause nerve cell toxicity, as shown in laboratory studies in animal models. It makes sense that riluzole works, and it is also logical to test other compounds in ALS that affect glutamate.

In the ten years that followed, no other drugs for ALS were found. Then, in 2006, investigators turned to a drug called ceftriaxone, which also affected glutamate. Identifying ceftriaxone as a potential drug for ALS stemmed from a program at the National Institute of Neurological Diseases and Stroke, where 1,040 compounds were screened to determine whether any might affect glutamate toxicity and, thus, be worthwhile to test in people. Drug development can take a long time, especially when a new compound requires safety testing in humans. The advantage of ceftriaxone was that it was already an FDA-approved antibiotic. Furthermore, in animal studies, ceftriaxone was found to slow disease progression, including in the SOD mouse model of ALS. Finally, ceftriaxone was believed to penetrate into the central nervous system, and, thus, it was felt to be a promising candidate for therapy.

The trial was carried out by the Northeast ALS Consortium, headed by Merit Cudkowicz. The ceftriaxone trial had a unique design aimed at moving quickly from Phase 1 through Phase 3. They used an adaptive trial design, which allowed the investigators to move from a Phase 1 dose-ranging stage, to a Phase 2 safety-data stage, and then directly into a Phase 3 trial. It was the first ALS trial to incorporate a Phase 1–3 adaptive design. Furthermore, because the drug had to be given intravenously and many of the patients were disabled, the investigators, for the first time, arranged for infusions to be given at home through ports that allowed IV access, something that is now done for edaravone, the second FDA-approved drug for ALS. It took six years, from 2006 until 2012, to go from Phase 1 to Phase 3.

The ceftriaxone trial enrolled 340 patients who received the ceftriaxone and 173 who received placebo; sixty-six people from the Phase 1 and Phase 2 trial participated in the Phase 3 trial. Unfortunately, the trial showed no effect on any of the outcome measures, although during the Phase 1 and 2 trials, the average ALSFRS-R disability score declined more slowly in those receiving the drug. This effect was not seen in the Phase 3 trial. The reason ceftriaxone didn't work, although it affected glutamate toxicity like riluzole, may have been because riluzole has a more complex mechanism of action that was not simply related to glutamate toxicity. Furthermore, the degree to which the ceftriaxone bound to its target in the nervous system was not clear. Unfortunately, the ceftriaxone trial is one of the many Phase 3 trials in ALS that showed promise in both animal models and Phase 1 and 2 trials but did not show an effect in Phase 3. Nonetheless, the ceftriaxone study introduced the novel adaptive trial design as a way of testing compounds that were already FDA approved, and the investigators showed that ceftriaxone penetrated the central nervous system, as measured by spinal fluid analysis.

Other trials of drugs directed at glutamate toxicity also failed in ALS—memantine, approved for the treatment of Alzheimer's disease, and another drug called talampanel. Because glutamate toxicity may play a role in progressive forms of MS, memantine has also been tested in MS, but it was not found to have significant benefit. In MS, the first drug approved was a beta interferon drug. Unlike drugs to target glutamate toxicity in ALS, subsequent trials of other beta interferon drugs in MS all showed positive effects, indicating that the target was valid. Paradoxically, although beta interferon was the first drug approved for MS, and there have been a number of trials showing beta interferons can help MS, as discussed earlier, the initial trial of beta interferon was tried for the wrong reason—namely, its antiviral effects, when it actually works through its immunologic effects.

It took twenty-two years after the FDA approved riluzole before a second drug, edaravone, was found to be effective in altering ALS progression and was approved by the FDA. Edaravone, marketed under the name Radicava, is an antioxidant and may act by affecting oxidative stress, although the drug company acknowledges that the mechanism by which it works in ALS is unknown. Oxidation is a chemical reaction that occurs in all cells and may produce toxic oxygen radicals that damage cells. When this happens, it causes oxidative stress, which antioxidants help control.

Antioxidants are used as food additives to protect against food spoilage. Vitamins such as vitamin C can also act as antioxidants. Oxidative stress is believed to contribute to a variety of diseases in addition to ALS, including Alzheimer's disease, Parkinson's disease, and cardiovascular disease. Oxidative stress occurs in a cell when there is a disturbance in the balance between the production of toxic oxygen species and antioxidant defenses. The *SOD1* gene is linked to oxidative stress in the cell.

One of the first reported trials of edaravone was a Phase 2 study published in 2006. Edaravone is a free radical scavenger that was approved for the treatment of acute cerebral infarction, or stroke, in Japan. Because oxidative stress is thought to be a major contributing factor in ALS, Japanese investigators, led by Hiide Yoshino, studied it in a small number of ALS patients. ALS patients have an increase of 3-nitrotyrosine (3NT), a marker of oxidative stress in the spinal fluid. Yoshino wanted to see what effect edaravone would have on 3NT in the spinal fluid. Earlier research had found edaravone protected neurons in so-called wobbler mice that have motor neuron loss and ALS-like symptoms, including muscle weakness and lack of coordination. The first human trial was small and open label; both patients and doctors were aware of the treatment. Twenty patients with ALS received either 30 or 60 mg of edaravone intravenously. Results suggested there was less decline on the ALS functional scale after treatment than in the six months prior to therapy. More importantly, the investigators found the level of 3NT in the spinal fluid was markedly reduced at the end of the sixth cycle of treatment. It appeared that the drug had hit its target.

However, a change in the course of a disease in an open-label study cannot be taken as evidence of a therapeutic effect, although it sometimes can give important clues that lead to controlled studies. The marked decrease in

MRI measures of inflammation in an open-label MS study led Al Sandrock to carry out controlled trials, which ultimately led to the approval of the drug Tecfidera for MS. Similarly, based on the open-label pilot trial of edaravone in ALS, investigators embarked on a larger, Phase 3 trial. They carried out a randomized, placebo-controlled trial that involved 102 subjects treated with edaravone and 104 treated with placebo. Unexpectedly, there was no effect.

When investigators conduct a trial that is negative and does not meet its primary endpoint, they carry out subgroup analysis with the hope of finding some people in the trial groups who responded well. If a subgroup is found that responded to treatment, a new trial must be done. Conducting subgroup analysis on a clinical outcome that wasn't the primary outcome measure initially identified for the trial does not count for the FDA in clinical trials. However, identifying subgroups can lead to the approval of a drug in a subsequent trial when it is shown that it works in the subcategory of patients identified in subgroup analysis. This process occurred in the approval of Ocrevus for progressive MS, which was based on subgroup analysis of a previous trial that showed younger people of a certain disease duration appeared to respond positively. Thus, the Phase 3 trial of Ocrevus that led to approval was done on this subpopulation. Of course, when the drug comes to market, all patients may be given the drug, although it was tested only in a particular subgroup.

In the case of edaravone, a second confirmatory Phase 3 trial was also negative. The investigators did not give up. They moved to a third confirmatory Phase 3 trial, where the inclusion criteria for the patients were narrowed even further. This time the trial was positive and published in *Lancet Neurology* in 2017. The randomized, double-blind trial enrolled sixty-nine patients to receive edaravone and sixty-eight to receive a placebo. The trial focused on early-stage ALS. This third Phase 3 trial was positive, and the primary outcome measure showed a decrease in the ALS disability score from 7.50 to 5.01 in the treated group, with a P value of 0.0013, clearly lower than the 0.05 cutoff.

Based on this and earlier studies, edaravone was approved by the FDA in the United States in 2017 for treating ALS. Although the authors wrote that there was no indication that edaravone would be effective in a wider population of ALS patients who did not meet the criteria of the patients in the trial, edaravone was approved for all patients with ALS. These results indicate how multiple trials and careful subgroup analysis may be required to find a treatment for a disease such as ALS, although one would hope drugs with wider efficacy will be found. The results of the riluzole trial and the edaravone trial show that both glutamate toxicity and oxidative stress appear to contribute to ALS. Unfortunately, two other drugs with anti-oxidative properties, coenzyme Q10 and creatine, failed, both in Phase 2 and Phase 3 ALS trials.

A great deal of interest in the ALS community surrounded creatine because of its oxidative effects and its potential effect on mitochondria, the part of the cell responsible for energy production. Creatine stabilizes mitochondria and is also important for energy or ATP production. Mitochondria serve as the cell's power plant, producing ATP, which cells convert into energy. Investigators found creatine is also involved in motor neuron function and prolonged motor neuron survival. Over a period of five years, three controlled, randomized trials of creatine were performed in ALS, involving a total of 386 participants who were given a dose of five to ten grams. No effect was seen in any of the trials. When the results of the first trial were negative, investigators had already begun the second trial and had initiated a third trial. In the third trial, investigators measured muscle strength as the outcome. However, no matter how the data were analyzed, there did not appear to be any effect, although creatine was used by ALS patients throughout the world and was being prescribed by neurologists. The reason for the trials' failure is unclear. It could relate to patient subgroups or dosing. More than likely, however, despite the theoretical basis for its use, it appears the drug simply did not work.

A provocative ALS study entitled "Lithium Delays Progression of Amyotrophic Lateral Sclerosis" was published by an Italian group in 2008 in the

Proceedings of the National Academy of Sciences. The authors reported that administering lithium to a mouse model of ALS resulted in protection of the neurons. Based on this, they immediately began treating ALS patients with lithium and reported that of the sixteen patients they treated, none had died during the fifteen months of follow-up. By comparison, 29 percent of a control group taking only riluzole died during the same period. Members of the control group were matched by age, disease duration, and sex.

The logic of using lithium made sense because animal and other studies suggested increased autophagy might help ALS, and lithium enhanced autophagy. Autophagy is a normal process in which damaged or redundant cells are destroyed and cleared. Increased autophagy would help clear abnormal proteins accumulating in the brain and spinal cord of ALS patients and is another way of protecting neurons. Needless to say, the report created a stir in the ALS community, and many ALS patients began taking lithium as an off-label treatment.

Clearly a formal trial needed to be done, and one was organized quickly with the National Institute of Neurological Disorders and Stroke and ALS societies providing funds outside the normal grant cycle. A unique design was put in place that involved rapid enrollment of patients, interim analysis, instituting clearly defined stopping rules, and allowing people who went on the placebo to move rapidly to lithium treatment should a positive effect be found. Participants would only be exposed to placebo for an average of seven months, and those with rapid progression could then be treated. As one can imagine, the study created a great deal of interest and enrolled quickly. Unexpectedly, no effect was found, and the study was stopped at the first interim analysis. A second study was subsequently published in 2012, involving 133 patients, and was designed to detect a larger effect than was projected in the first controlled trial. Again, nothing was found. Following the first study, scientists could not consistently replicate the initial findings in animals.

Because so many ALS patients began taking lithium independent of a formal clinical trial, ALS patients participating at the website

PatientsLikeMe built a lithium-specific data collection tool. Using self-reported patient data collected online is an example of another way to investigate the effect of lithium. Approximately half of ALS patients take vitamins and unproven treatments. As with previous studies with the antibiotic minocycline, patients rushed to take lithium. As it turns out, the PatientsLikeMe cohort was also unable to find an effect of lithium use.

As with other failed trials, it is not known why no effect with lithium was found, but almost a decade after the unsuccessful lithium trials, investigators reexamined the failed trials by combining data from the three randomized lithium trials for all 518 patients. They stratified patients genetically to determine whether any genetic traits could predict why some patients did not respond to lithium. They looked at the *C9orf72* gene, as well as carriers of *UNC13A*, a gene that conveys risk for sporadic ALS. They found the twelve-month survival for those with the *UNC13A* genotype improved from 40 percent to 70 percent. As discussed previously, post-hoc analysis can sometimes give clues to discover responsive subgroups. This was one of the first genetic post-hoc analyses in ALS. These findings have implications for moving ALS into the realm of precision medicine and for the design of further trials. This approach is analogous to what is routinely done in the treatment of breast cancer, where the chosen therapy and response to therapy is linked to genetic subtypes. Hopefully, the failed results in the lithium trials will lead to successful ALS trials in the future based on precision medicine and genetic subtyping.

Just as the negative results of the lithium trials were being reported, a groundbreaking paper was published in 2011 in the highly regarded journal *Nature Medicine* on the effects of dexpramipexole in patients with ALS. The study was led by Merit Cudkowicz. Dexpramipexole was one of the most exciting drugs to be tested in ALS at that time. In 2006, a paper published by a group in Germany showed that a form of dexpramipexole had properties that could be of benefit in ALS by targeting mitochondria, the energy-producing machinery of the cell, and by affecting toxic molecules in the cell that caused damage to neurons. In addition, the drug ameliorated disease in the SOD mouse model of ALS and penetrated nicely into the brain. Other investigators expanded these findings and showed beneficial effects of dexpramipexole on neurons in cell culture assays.

These encouraging properties of the drug led to an open-label study with thirty subjects, which showed a slowing of functional decline in ALS. Based on this, 102 subjects at twenty US sites received placebo or increasing doses of dexpramipexole. Investigators found that the drug was well tolerated, and there were clinically significant results in the group receiving 300 milligrams with a P value of 0.046. The study was double blind and had a unique design involving a first phase in which the drug was given for twelve weeks and a second phase after the drug was discontinued for four weeks. This unique study design allowed the same cohort of subjects to participate in two separate studies of the drug. The authors were very encouraged by the results and wrote, "To our knowledge no other drug has shown a clinically significant effect on the decline of the ALSFRS-R in a properly controlled clinical trial, and no other study has shown effects on both function and mortality."

The positive results published in *Nature Medicine* led to a Phase 3 trial, called EMPOWER, which was funded by Biogen. Al Sandrock told me Biogen first heard about dexpramipexole from Knopp Biosciences before the Phase 2 trial, having seen results of the drug on the SOD mouse. Sandrock told Knopp to talk to Merit Cudkowicz and do a Phase 2 study. "If it's positive," Sandrock said, "come back to Biogen." They came back after the positive trial, and Sandrock was encouraged that there was a dose-dependent slowing of progression. He reasoned that if there was a dose effect, then the drug was having real biologic effects. Biogen was never 100 percent convinced of the proposed mechanism of action and felt that the effect of the drug in the animal model was only modest. Nonetheless, Biogen paid Knopp $50M for the opportunity to fund it and spent close to $100M on the trial. The Phase 3 trial enrolled 943 patients. Sandrock said it was one of the most rapidly recruiting trials Biogen ever had. They enrolled more patients than they needed because so many people wanted to participate. The 953 participants were enrolled over a six-month period between March 28, 2011 and September 30, 2011. Sadly and unexpectedly, when the Phase 3 trial was unblinded, no effect was seen. There was no change in the ALSFRS-R score, a decrease of 13.34 in the treated group versus an almost identical decrease of 13.46 in the placebo group. Time to death was also no different in the groups. This result was especially disappointing because the EMPOWER trial was considered the best hope for ALS at the time.

Prior to joining Biogen, Sandrock had trained at Massachusetts General Hospital and had cared for ALS patients with Bob Brown. He witnessed the disease firsthand. He told me that when he heard of the negative results of the EMPOWER trial, it was one of the worst days of his life, similar to when he learned that the MS drug Tysabri caused the fatal brain infection PML. Luckily, it was around the holidays, and he was able to take long walks. He was never criticized by management or by his CEO for the failed trial. They understood how important the trial was and how bad a disease ALS was.

Fortunately, Biogen has not given up on ALS, and the company is now conducting specifically targeted trials, doing gene therapy for patients with genetic mutations in *SOD* and *C9orf72*, which have initial promising clinical results—see below. Sandrock told me that they came up with a novel outcome measure from the dexpramipexole database: a quantitative muscle strength measure based on a handheld dynamometer, a sophisticated device that is placed on the muscle to measure muscle strength rather than simply having the patient push against the doctor's hand. Again, something was learned from a failed study.

In discussing the failed EMPOWER trial, Cudkowicz and her colleagues wondered whether investigators should perform several Phase 2 studies to make sure the clinical effect was real. In addition, a specific biomarker linked to response to therapy, such as an MRI measure of brain inflammation used for relapsing MS, would be a major advance.

As they had with dexpramipexole, large trials of other neuroprotective drugs also failed, including a trial of TCH346, which blocks apoptosis, or cell death. Some 591 patients were enrolled in forty-two sites in Europe and North America. The therapeutic rationale was strong, and there were impressive preclinical data in animals, as well as a hint of clinical efficacy in a small screening trial. Unfortunately, TCH346 did not pay off. Olesoxime is a drug that was identified in a screen for compounds that protect neurons in the test tube and showed positive results in the animal SOD model but failed in an eighteen-month, double-blind, placebo-controlled trial involving 512 patients. A trial of erythropoietin, which is a growth factor for red blood cells, was found to have neuroprotective effects on motor neurons, but it also failed in a Phase 3 study. Several neuroprotective strategies that worked in the test tube or in animals did not work in human subjects with ALS because apparently the disease is too complex to be resolved with one drug treatment, as it has multiple factors driving it.

Neuroprotective drugs protect the neuron from being damaged by toxic factors. Another approach for treating ALS is to provide growth factors, which theoretically can help repair or grow damaged motor neurons. There are a number of growth factors that help maintain neurons, including brain-derived growth factor, ciliary neurotrophic factor, and insulin-like growth factor. However, despite the appeal of these neurotrophic factors, they have all been tried in ALS without much benefit.

The growth factor with the most extensive testing is insulin-like growth factor. In preclinical studies, insulin-like growth factor promoted the survival of both spinal and facial motor neurons in models of nerve transection. It also had some positive effects in the wobbler ALS mouse model. Because patients with ALS had more insulin-like growth factor receptors in the spinal cord than people without ALS, it could be a target of therapy. An initial double-blind, placebo-controlled trial in 1997 of 266 patients showed that progression of functional impairment was 26 percent slower in patients receiving insulin-like growth factor than in those receiving placebo, with a P value of 0.01. However, in a second trial a year later, no effect was seen in the 183 patients treated. Ten years later, another trial was undertaken to address the inconsistencies. A total of 330 patients from twenty medical centers were given insulin-like growth factor, measuring their survival and muscle function and strength. Unfortunately, this trial was negative. Other large trials of growth factors were also negative, including a 1996 trial of human ciliary neurotrophic factor involving 730 patients, and a 1999 trial of brain-derived neurotrophic factor involving 1,135 patients. The results from all these trials suggest that growth factors are not strong enough to rescue motor neurons from the catastrophic effects of ALS.

A recent neuroprotective treatment that has met with success in ALS is a compound called AMX0035, which is a combination of sodium phenylbutyrate and taurursodiol. Taurursodiol is an over-the-counter supplement that is a natural bile component and affects endoplasmic reticulum dysfunction in the cell. Sodium phenylbutyrate is a medication that affects mitochondria and is used for a pediatric urea disorder. Dysfunction of both the endoplasmic reticulum and the mitochondria has been implicated in cellular dysfunction in ALS.

The idea for the combination came in 2013 when two Brown University college students, Joshua Cohen, a biomedical engineering major, and Justin Klee, a neuroscience major, became interested in ways to treat brain disease by affecting neuronal death. A search of the literature showed that these two compounds had shown positive effects in models of both ALS and Alzheimer's disease. Sodium phenylbutyrate had actually been tested in ALS by Merit Cudkowicz in a pilot trial. The students hypothesized that if the two were combined, there would be a beneficial effect in neurological disease.

They approached Rudi Tanzi, an Alzheimer's disease expert at Mass General, who suggested they test their hypothesis on neurons in culture. The results were dramatic. Unable to find someone interested in testing their protocol in Alzheimer's disease, they turned to ALS. They named the combination AMX0035 and formed a company called Amylx Pharmaceuticals. The pieces were finally in place to conduct a trial.

The trial was called CENTAUR and was the first clinical trial in ALS to be supported by funds raised by the ALS Ice Bucket Challenge, in addition to the ALS Association. The trial was a Phase 2, multicenter, randomized, placebo-controlled trial performed at six sites in the United States. Enrolled subjects had ALS symptoms within eighteen months of starting treatment. They were allowed concomitant treatment with edaravone and riluzole. They received a fixed-dose combination of sodium phenylbutyrate and taurursodiol, in a powder that was dissolved in water at room temperature. It was swallowed or given through feeding tube once a day for three weeks and then twice a day for a total of twenty-four weeks. An identical-looking placebo powder that had a bitter taste similar to the drug combination was given.

The primary outcome measure was the rate of decline on the ALSFRS-R functional rating scale. Secondary outcome measures included muscle strength, breathing, hospitalization, time to death, and a measurement of blood neurofilament levels to assess neuronal damage. Randomized in a two to one ratio, eighty-nine patients received the drug combination and forty-eight received placebo. Excitingly, the results on the primary outcome measure were positive and showed a slowing of functional decline. The mean decline in the ALSFRS-R score was -1.24 points per month on the drug combination and -1.66 points per month on the placebo, with a $P = 0.03$. Trends, but not statistically significant differences, were observed in the secondary measures of breathing, muscle strength, and hospitalization. No changes were observed in the blood biomarkers of neurodegeneration.

The CENTAUR study was published in the *New England Journal of Medicine* in September 2020 with an accompanying editorial entitled "Incremental Gains in the Battle Against ALS." Although the clinical effects were modest and 19 percent of those receiving AMX0035 discontinued treatment because of side effects, there have been few pharmacological agents that have been shown to affect functional decline in ALS. It is clear that larger trials and longer-term follow-up are needed. A Phase 3 trial of taurursodiol (taurourodeocholic acid) alone is ongoing. Because both taurursodiol and sodium phenylbutyrate are available as individual agents, it may be difficult to perform a Phase 3 trial. Some feel AMX0035 should be approved by the FDA based on the Phase 2 results. Although not a home run, the trial was a much-needed success in the face of so many trials that have failed. The CENTAUR trial showed that drugs targeting cellular mechanisms that affect endoplasmic reticulum and mitochondria can have an ameliorating effect on ALS. AMX035 may one day be part of a cocktail of combination therapy for ALS, an approach that has been so successful in cancer.

ALS AND INFLAMMATION

Another approach that investigators have taken has been to use anti-inflammatory drugs to treat ALS. Although defects in cell metabolism in the motor neuron are central to ALS, inflammation may also play a role. Targets of therapy include immune cells in the brain (microglia and astrocytes) that surround the motor neuron and immune cells from the blood (monocytes) that infiltrate the brain.

As discussed, we and others are trying to find ways to affect microglia and monocytes to treat ALS. In fact, the anti-inflammatory drug glatiramer acetate, approved for treating relapsing-remitting MS, showed some promise in early ALS animal studies, but it had no effect in a double-blind, multicenter, placebo-controlled trial in ALS. Minocycline is an antibiotic that appears to affect microglial cells and had neuroprotective effects in animal models of stroke and ischemic injury. A trial of minocycline given to people with early MS showed positive results. However, a double-blind, Phase 3, placebo-controlled trial in ALS showed that not only did the drug not work but it might have even been harmful to patients, who deteriorated faster on the ALS functional score. As discussed previously, pioglitazone, an oral, antidiabetic drug that has anti-inflammatory properties, was tested in Alzheimer's disease based on epidemiology data, without positive results. Pioglitazone was also tested as a potential therapy for ALS based on positive effects in ALS mouse models, associated with decreased levels of inflammation. However, a double-blind trial of 219 patients with ALS showed no effect with pioglitazone. Celecoxib is an FDA-approved drug for arthritis that blocks prostaglandin synthesis. Prostaglandins are involved in the immune response. Celecoxib also showed some benefits in preclinical testing but failed in a double-blind study of 300 subjects with ALS.

Despite these negative findings, a recent study suggests that an anti-inflammatory drug that affects monocyte activation may be helpful in ALS. Neuraltus Pharmaceuticals is developing a drug called NP001. In initial studies, it caused a dose-dependent reduction in measures of monocyte activation and also had an effect on inflammatory monocytes in ALS blood. This is consistent with our studies in which we found activation of monocytes in the blood of ALS patients similar to that of patients with MS. A randomized Phase 2 trial of NP001 demonstrated positive effects, including slowing of disease progression, especially among those who received a high dose of therapy and those who had greater inflammation in the blood as measured by C-reactive protein. Additionally, most of the responders had elevated levels of the inflammatory marker interleukin 18 and were positive for lipopolysaccharides, molecules that elicit a strong immune response. Both decreased after treatment. These results suggest that a subset of patients with measurable inflammation may respond positively to the drug. Sadly, a confirmatory, double-blind, placebo-controlled, Phase 2 study of NP001, involving 138 ALS patients, showed no differences in the rate of functional decline or breathing capacity.

Another anti-inflammatory drug that is being tested is masitinib, which is being developed by AB Science. Masitinib targets mast cells, which are involved in allergic responses and may affect other cells that play an important role in ALS, such as microglia. In a double-blind, placebo-controlled,

Phase 2-3 study, 191 patients were treated with masitinib for forty-eight weeks. The company reported the ALS functional rating score showed a positive effect with a P value of less than 0.01. Given its general anti-inflammatory effects, masitinib has also been tested in patients with progressive MS and Alzheimer's disease, with positive effects reported in both diseases. The drug is also being studied in various types of cancer and other inflammatory diseases.

It is unclear how a drug can affect so many different diseases, but it may be that inflammation is a secondary amplifying factor. Independent of these theoretical considerations, the bottom line will be clinical results and whether masitinib shows positive results in Phase 3 trials. As discussed, many drugs that appeared promising with early positive results did not ultimately succeed in ALS.

ARE STEM CELLS THE ANSWER?

My patients always ask me about stem cells. The hope is that stem cells may be able to rebuild or substitute for damaged cells in the nervous system, whether for ALS, MS, Alzheimer's disease, or Parkinson's disease. Stem cell trials in ALS and MS are being undertaken. To see why stem cells are an avenue of inquiry, it is important to understand what stem cells are, the different types of stem cells, and how they are being used both for research and treatment.

Embryonic stem cells come from an embryo and have the ability to develop into the different cells in the body. Fetal stem cells harvested from fetal tissue can also take on different forms. Stem cells also exist in adults. In fact, there are small numbers of stem cells in many parts of the body from which new cells may derive. For example, there are small numbers of stem cells in the brain that can give rise to cells like microglial cells.

The use of stem cells in medical research was initially plagued by ethical concerns over the use of fetal or embryonic tissue. This is no longer an issue. As discussed, scientists are now able to take a skin biopsy or a blood sample from a person and turn these differentiated cells into an undifferentiated stem cell that can then be reprogrammed into any type of cell. These cells are called iPSC cells, or induced pluripotent stem cells. In 2012, the Nobel Prize in Medicine was awarded to Shinya Yamanaka, who demonstrated that by inserting only four genes into a mature adult cell, he could turn it into a stem cell.

Given these advances in the laboratory, it is only logical for a patient to ask, "Why can't you give me some stem cells to replace my damaged motor neurons?" Unfortunately, we are far from making motor neurons in the test tube and transplanting them into people. In fact, we can't yet do it in mice. For now, induced stem cells are being used for research and screening of new drugs. Induced stem cells made from patients with MS or ALS or Parkinson's are created in the laboratory, and these cells contain all the genetic material of the patient. This allows experiments to be conducted to understand the genes that may be involved in the disease, or to use the stem cells to screen for new drugs. Research centers such as ours derive stem cells from patients, and the ALS Association has helped support central facilities to create stem cells from ALS patients that can then be used by all investigators.

A frequently used type of stem cell is mesenchymal stem cells; these are being explored as a form of treatment in ALS, MS, and other neurological diseases. This type of stem cell is usually harvested from a person's own bone marrow, grown in a tissue culture dish, and then reinjected into the patient's spinal canal as treatment. One theory of their mode of action is that they release neuron-protecting factors that help protect or improve the function of dying motor neurons in an ALS patient. These stem cells can be programmed in such a way that they produce specific neuron-protecting factors in the test tube. They may also act to dampen the inflammation in the nervous system of ALS patients. The FDA has approved a number of these stem cell programs based on technology that is now available and positive results have been shown using these stem cell therapy approaches in some animal models.

A group in Michigan has been investigating a stem cell line derived from the human spinal cord, which is injected in the cervical spinal cord of patients with ALS. The investigators hypothesized that these stem cells may be most effective when injected in the cervical and lumbar regions of the spine, as this would help in muscles associated with breathing. Technical aspects of the study had been worked out, including a reproducible surgical procedure allowing the injection of the stem cells into the cervical cord, which could be replicated in multiple centers. Clinical results comparing the rates of decline in patients receiving cervical stem cells to historical controlled groups showed no differences in the mean rates of clinical progression. However, the initial goal of the study was safety and technical feasibility.

A group in Israel has been investigating the safety and clinical effects of mesenchymal stem cells, which secrete neurotrophic factors, given both to ALS and progressive MS patients. They call their stem cells MSC-NTF cells. In comparing the clinical course of patients before and after injection of stem cells, there was some suggestion of stabilization for a period of six months. The numbers, however, are very small. A group in the Czech Republic has also treated patients with stem cells injected into the spinal cord. They reported that there might have been a decrease in the progression of disease in some patients.

A group at Cedars-Sinai in Los Angeles has taken a different approach, in which they give spinal cord injections of stem cells designed to be glial cells such as astrocytes to help support the motor neuron. As described earlier, microglial cells and astrocytes can be toxic in ALS, and this trial is aimed at providing healthy astrocytes in patients' spinal cords.

Companies have been formed to develop stem cell treatment, including Neuralstem, BrainStorm Cell Therapeutics, and Q Therapeutics. BrainStorm is developing treatment with a stem cell preparation called NurOwn. It showed positive effects after spinal canal injection in a Phase 2 trial, but a Phase 3 trial was not successful, though there may have been a response in a subgroup of patients. Stem cell centers have been established in South Korea, Thailand, and Germany, where patients can pay for stem cell therapy. The center in South Korea has found clinical improvement in a controlled trial in which patients receive two injections of mesenchymal stem cells into their spinal canal, but there is no effect on overall survival. Their treatment has been provisionally approved by the Korean equivalent to the FDA.

Cochrane Reviews, which provides comprehensive reviews of treatments in neurological and other diseases, identified twenty-five published trials involving 680 people who were treated with stem cell therapy. There were no double-blind or placebo-controlled trials. Although there were some complications, the treatments were generally safe. Randomized, controlled trials are underway that will determine the degree to which there is efficacy. Stem cell therapy is still in its early stages, and important questions need to be answered, including how often treatment must be given, which patients, if any, respond, and long-term effects.

GENE THERAPY

The discovery of the SOD mutation in 1993 by Bob Brown was a seminal event in the history of ALS. It created a mouse model that could be studied and enabled identification of potential mechanisms in the disease. It has also provided an avenue for the development of new therapies. Although the discovery of the SOD mutation initially led investigators to believe that ALS patients simply needed to be given the SOD protein to make up for what they were lacking, it has now become clear that the SOD mutation does not cause the loss of a vital molecule that needs to be replaced. Instead, the genetic abnormality causes the appearance of a molecule or series of molecules that damage motor neuron function.

When investigators speak about disease and potential targets, they refer to upstream and downstream events. A disease such as ALS begins like a small stream of water that grows until it becomes a river and then a flood. The small stream of water is upstream, the flood is downstream. A similar analogy is that of an avalanche, in which a small disturbance in the snow upstream initiates a damaging avalanche downstream. Diseases are easiest to treat and ultimately cure if the inciting event can be identified and treatments can occur as far upstream as possible. It is possible that so many treatments haven't worked in ALS because we are treating downstream events, and the trigger for the avalanche has already occurred.

This is also the situation in Alzheimer's disease, where it is likely that by the time patients have problems with cognition, the stream has swollen to a flood. In ALS, in those with the SOD mutation, the SOD gene is the most upstream event possible—identification of a specific gene that initiates the disease. If it were possible to correct that genetic defect, there would be an excellent chance of treating the disease in this small group of patients, perhaps even curing the disease. As discussed below, this concept is now being tested in clinical trials of ALS patients with the SOD mutation, and it has met with initial success.

ALS is a terrible disease. The average time of survival from onset to death is between three and five years. If a treatment worked, complicated measures of function to test efficacy wouldn't be needed. Like cancer, one could simply measure survivors versus deaths.

Indeed, the outcome measure everyone would hope for in ALS—survival—occurred in a trial of gene therapy treatment for spinal muscular atrophy, where dramatic success was achieved. Spinal muscular atrophy, a fatal motor neuron disease, is one of the most prevalent and devastating disorders in childhood. It is caused by the mutation of a single gene, *SM1*, which results in the loss of production of the SMN protein, which is required for normal muscle function. A double-blind, randomized, sham-controlled gene therapy trial for spinal muscular atrophy was stopped early because many of the infants treated survived, whereas none survived in the control group. This result led to FDA approval.

The gene therapy used in spinal muscular atrophy involved antisense oligonucleotides (ASOs). How do antisense oligonucleotides work? Remember the pathway by which our genes work. Genes are made of DNA, and the DNA is transcribed into messenger RNA. The messenger RNA is then translated into a protein, which carries out normal function. A defective gene results in an abnormal messenger RNA, which leads either to an abnormal protein or the lack of a normal protein. Thus, when the structure of a gene is known, it is possible to synthesize an analogue such as an ASO that can bind to the messenger RNA and, in effect, shut off a gene.

The potential for the therapeutic use of ASOs has been recognized since the 1970s. However, decades of research were required to understand their basic function, build a structure that worked in people, and allow clinical testing. Preclinical studies showed that an ASO could cure mouse models of spinal muscular atrophy. These studies were quickly followed by an open-label, Phase 1 clinical trial where different doses of the ASO were injected into the spinal fluid. With safety and route of administration established, a Phase 2, open-label, dose-escalation study in twenty infants was performed and showed improvement of motor function. Finally, a Phase 3 clinical trial was performed with an ASO called nusinersen in infantile onset spinal muscular atrophy. The ENDEAR study included 173 infants; 121 were part of a double-blind, sham-controlled investigation. Interim analysis showed nusinersen reduced the risk of death or permanent ventilation by 47 percent in the eighty-two infants receiving the treatment. This led to FDA approval of nusinersen, which is marketed under the name Spinraza. A gene therapy approach has also been successful in another inherited muscle disease of children, called Duchenne's muscular dystrophy, in which there is a defect in production of the dystrophin protein.

ASOs have opened a potential way to treat SOD familial ALS. An ASO for the SOD mutation would be designed to decrease the toxic SOD1 protein. Studies in mice have been successful, and a Phase 1 first-in-human study was carried out by Timothy Miller at the University of Washington in St. Louis and Merit Cudkowicz. It was published in 2013. The ASO was injected into the spinal fluid over a dose range to determine feasibility. Apart from headaches, there were no safety concerns. However, because of the low concentration of the ASO used, the study did not show a reduction of the toxic SOD1 protein in patients.

A second-generation, SOD1-targeted ASO compound called tofersen was designed by Biogen. Promising results of a Phase 1-2 clinical trial in forty-eight patients with SOD1-ALS were reported in the *New England Journal of Medicine* in July 2020, along with a report of two patients treated with an adeno-associated virus that also targeted SOD1. An editorial accompanying the articles heralded "The Beginning of Genomic Therapies for ALS." It was a true milestone on the road to find a cure for genetic forms of ALS.

The Biogen sponsored study of tofersen enrolled patients at eighteen different sites. In a double-blind study, patients received placebo or five doses of tofersen (20, 40, 60, or 100 mg) given into the spinal canal over twelve weeks. The primary outcome measure was safety and whether the treatment lowered SOD1 levels in the spinal fluid, which would demonstrate that the drug had indeed affected its target. Neurofilament levels were measured in the spinal fluid as a measure of damage to neurons in the spinal cord. The results showed that in the patients given the highest dose, there were lowered SOD1 protein levels in the spinal fluid and less nervous system damage as measured by neurofilament levels. Although an exploratory outcome, clinical decline as measured by the ALSFRS-R, respiratory function, and muscle strength was less in those who received the highest dose of the ASO. The major side effects were related to the spinal tap that was done to administer the treatment. Based on these results, a Phase 3, randomized, placebo-controlled, double-blind trial is underway, and Al Sandrock at Biogen is hopeful that the tofersen trial will succeed where their trial of dexpramipexole failed.

In the same issue of the *New England Journal of Medicine* containing the Biogen tofersen study, a team led by Bob Brown reported a gene therapy approach in which a vehicle made from a nontoxic adenovirus was used to deliver a microRNA to silence the *SOD1* gene. Two patients with familial SOD1 ALS were treated with a single spinal canal injection. One was a twenty-four-year-old man whose mother had died of ALS at age forty-five, and the other, a fifty-six-year-old man with a family history of ALS. The first patient succumbed to the disease, but at autopsy the SOD1 levels in spinal cord tissue were decreased, suggesting an effect of the treatment. Inflammation caused by the viral vector required the use of immunosuppressive drugs in the second patient. It appears that gene therapy targeting the *SOD* gene will ultimately be successful and will come full circle from Bob Brown's discovery of the *SOD* gene in 1994 to a treatment that silences the gene and treats ALS.

Although *SOD1* accounts for only a small percentage of inherited ALS, *C9orf72* is the most prevalent genetic mutation, accounting for approximately 40 percent of all inherited forms. A proposed mechanism of action includes the loss of the C9orf72 protein function and perhaps toxic RNA species. A trial is now underway using ASO gene therapy directed against *C9orf72* for the treatment of ALS and the genetic modifier ATXN2, which modulates TDP-43 expression in ALS. It is not known the degree to which sporadic ALS may also relate to these genes, and it is possible that ASOs for genetic mutations linked to familial ALS could potentially have even wider applications in sporadic forms of ALS. Although treating patients may be a number of years away, the exciting point is that clinical trials of gene therapy are underway.

As discussed, we found that the inflammatory microRNA miR-155 is elevated both in the microglia and blood monocytes of patients with ALS. Plans are underway to test gene therapy by injecting anti-miR-155 into the spinal fluid, which would quiet overactive microglial cells. This is more downstream than affecting SOD protein production but could positively affect sporadic cases and potentially be useful in other diseases with activated microglia, such as MS and Alzheimer's disease.

Investigators are also using other approaches to target *SOD1* in ALS in an attempt to quiet the toxic SOD1 protein. Arimoclomol is a drug that affects protein folding. It promotes natural folding of nascent proteins and the refolding of misfolded proteins and could help repair the misfolded SOD1 protein. Arimoclomol was shown to work in animal models, and a Phase 1-2, double blind, placebo-controlled trial in patients with rapidly progressive SOD1 familial ALS was performed on thirty-eight patients. Although the numbers were small, there's a suggestion that the patients treated with arimoclomol did better than the control group, and a dose of 200 mg three times a day was tolerated for twelve months. A Phase 3, randomized, placebo-controlled trial in ALS is now underway.

Because families with SOD1 familial ALS can be identified, Oleg Butovsky at our center is leading a group to study white blood cells in these patients. This study may permit us to identify asymptomatic patients whose disease process has already begun, allowing for treatment as early as possible. Furthermore, understanding the natural history of SOD1 familial ALS and changes that can be observed on brain and cervical cord MRI imaging will lead to more targeted approaches to therapy.

CURING ALS

Although we have learned a great deal about ALS in the last number of years, the central question remains: How will we find an effective treatment and ultimately a cure? Are we on the right track? Is there a missing piece? Why have there been so many negative Phase 3 clinical trials in ALS? The answers lie in a number of directions. It is possible that the primary mechanism behind motor neuron degeneration in ALS is not yet known, and until we find that exact mechanism, we will not have a treatment. I do not believe this is the case. When I spoke with leading ALS researchers about how ALS would be cured a hundred years from now, they believed the factors that have been identified to date make biologic sense: glutamate toxicity, neurotrophic factors, oxidative stress, mitochondrial dysfunction, protein misfolding, abnormalities in handling cellular debris, and inflammation. Treatment is a matter of being able to target these contributors to

ALS in an effective manner. Gene therapy for those with *SOD1* or *C9orf72* mutations may provide dramatic results.

One of the reasons for the failed trials may be that the treatment was begun too late, at a point when too much happened downstream and the tide couldn't be turned. The avalanche gathered too much snow. With better clinical trial design and more sophisticated biomarkers, testing new drugs will be more efficient. Because there are subcategories of ALS, different disease processes may be important in different subgroups. The genetic responders in the lithium trial may provide a clue to identifying these subgroups. It may be that ALS will be cured piece by piece in subgroup by subgroup.

A key component of ALS is that not only motor neurons contribute to ALS; supporting cells such as the microglia, astrocytes, and oligodendrocytes also contribute to the disease. This offers hope in that our ability to target these cells will help all types of ALS, which will help slow the avalanche. We may need to target more than one factor at once.

Cancer therapy, for example, requires combinations of treatment, and one would expect that combination therapy will be ultimately required to control ALS. As discussed, Amylyx is conducting a trial that uses a combination of an antioxidant and neuroprotective compound called TUCA, and sodium phenylbutyrate, which promotes motor neuron cell survival. In animals, investigators are cotargeting independent pathogenic mechanisms and are obtaining better results. For example, they are combining suppression of microglial inflammation with reduction of SOD in astrocytes and motor neurons.

There are certain areas that have not yet been targeted in ALS that one day may become part of an effective treatment. One of these areas is the microbiome, which is abnormal in MS. It is also being evaluated in Alzheimer's disease and, as we will see, appears to play a major role both in Parkinson's disease and cancer. Initial studies suggest that butyrate-producing bacteria are deficient in the gut, and increasing them may benefit those with ALS. A major discovery from a group at the Weizmann Institute emphasizes the potential importance of the microbiome in ALS. In animal studies, they found that disrupting the microbiome in SOD1 mice by treating with oral antibiotics caused worse disease in which animals did not live as long. Evidently, the antibiotics removed bacteria that have a protective role in the disease. Furthermore, the investigators found a decrease in a bacteria called *Akkermansia* in animals with worse disease, and administering *Akkermansia* improved mouse survival. The ameliorating effects of giving *Akkermansia* was related to the production of certain molecules, including nicotinomide. These findings related to *Akkermansia* and nicotinomide were found in ALS patients as well. We have also found that antibiotics make disease in the SOD mouse worse; that the gut microbiome can affect microglial cells in the brain; and that a molecule that makes *Akkermansia* grow in the gut improves ALS in animal models. We have also identified specific bacteria that are beneficial when given to animals. These results suggest that modulating the gut with bacteria that strengthen the immune system or provide neuroprotective factors could be of benefit in ALS.

As discussed, in addition to the *SOD* gene, there are other genetic mutations linked to ALS, the most common being in the *C9orf72* gene. However, when researchers created the C9orf72 animal model of ALS, they were surprised to find that unlike SOD mice, C9orf72 mice did not develop symptoms of ALS. However, as other investigators began to study the mice, a group in the Harvard Institutes of Medicine building found that some mice in their facility became paralyzed. This was in stark contrast to mice housed across the river at the Broad Institute in Cambridge. Those mice did not get sick.

Because the mice were genetically identical, the researchers reasoned that it had to be something in the environment, and they hypothesized that the housing conditions led to a different microbiome, which in turn affected the development of ALS. There was an easy experiment to test this hypothesis. The researchers swapped the microbiome between the two different animal facilities and found that the mice in Cambridge became sick if they received microbes by fecal transplant from the Boston facility, whereas the healthy animals in the Boston facility did not get sick if they received microbes from the Cambridge facility. Their task now is determining which microbes are responsible for the difference. This shows that having genetic risk is not enough; the gut microbiome is needed to trigger ALS in a genetically susceptible host.

Given the strong connection between the gut and the brain, targeting the microbiome is a new avenue of ALS therapy, although we do not believe that the gut is the primary driver of the disease. Trials of fecal microbiota transplantation and special probiotic mixtures are being considered for ALS.

Although certain environmental factors have been identified, no strong environmental factor appears to be driving ALS. Nonetheless, the environment can play an important role. Identical twins are more likely to both have ALS than fraternal twins, but that is not always the case. Also, identical twins who both have ALS often don't have the same expression or progression of the disease. This is because the same environment is experienced differently by genetically identical individuals. Interestingly, even genetically identical mice in the same cage may have differences in their brains and their immune systems.

When I join Merit Cudkowicz at the Healey Center for ALS research at Mass General Hospital, I have a feeling of hope, given recent positive trials and many new leads. I'm struck by the courage of those with the disease, and the dedication and determination of people caring for them. Although there are only three FDA-approved drugs for ALS, we are better able to care for our patients today than in the past. Multidisciplinary clinics provide comprehensive support, including diet, help with speech therapy, walking, and mobility aids. The computer has enabled ALS patients to remain connected to the world, and technology can give them a voice when they can no longer speak. Stephen Hawking continued his work throughout his life, his ALS notwithstanding. In Boston, at Leonard Florence House, with help from the philanthropist Miriam Adelson, there is an ALS facility that provides support for activities of daily living so that patients achieve a certain degree of independence.

In the summer of 2014, something unprecedented happened that helped to change the face of ALS research. Called the Ice Bucket Challenge, it featured videos—many of which went viral—of celebrities such as LeBron James and Bill Gates dousing themselves with ice water and challenging others to do likewise in the name of ALS. It was unlike anything we've ever seen in fund-raising for medical research. I joined the thousands who took part in the Ice Bucket Challenge, which not only raised awareness about the disease but also raised close to $200 million. The Ice Bucket Challenge gave ALS research an enormous infusion of funding, and Bob Brown told me the money has made a huge impact. He said that more money for research will mean a cure will come faster for ALS. The Ice Bucket Challenge helped fund the successful trial of the neuroprotective agent AMX0035 discussed earlier. The Director's Office at the NIH recently put out a call for Transformative Grants in ALS, similar to what I was awarded in Alzheimer's disease.

What will the answer be? As I see it, gene therapy for specific subsets of patients; biomarkers that identify people early so one identifies and treats the disease upstream before the avalanche occurs; combination therapy that targets different pathways; and tamping down the immune system so it doesn't amplify the disease process. Ultimately, all patients will be genotyped, and halting disease progression will be the first achievable goal. From the clinical trial standpoint, Merit Cudkowicz has created a master platform trial approach that will test more than one drug at a time, share control groups, and develop biomarkers in an adaptive trial approach that has been successful in oncology. In addition, there will be small groups with precision treatment directed at specific pathways, such as a trial of the anti-epileptic drug retigabine, which targets electrical hyperexcitability in ALS patients. There are some suggestions that people who are more athletic and more fit may have an increased risk for ALS, a cruel irony for a paralytic disease.

The brain, of course, is the seat of our consciousness. I confront life's big questions whenever I see a patient with ALS, and I think of Tanya, whom we were unable to help.

Because of the rather rapid progression of the disease, ALS forces patients and doctors to confront and question life in a more acute way than we would ordinarily, as exemplified by Brandeis professor Morrie Schwartz, who had ALS and described it as eventually leaving "your soul, perfectly awake, imprisoned inside a limp husk." Schwartz's interviews with former student Mitch Albom became the subject of a book, which also became a movie, *Tuesdays with Morrie,* an exploration of life and living in the face of mortality.

All the doctors I interviewed for this book find solace in caring for people with ALS. They overcome feelings of hopelessness by working for a cure. Bob Brown, his hair now gray since the time he began in my lab more than thirty years ago, told me that many of his ALS patients, after the shock of diagnosis, get their bearings and go back to being themselves and being with their families for the time they have left. There are many stories of heroic and determined patients with ALS who have somehow overcome their disability.

Unlike Alzheimer's disease, most patients with ALS retain their cognition. Thus, Stephen Hawking lived a full and productive intellectual life despite his illness. There are countless examples of those who have not given up. For example, there is the story of a young Irishman named Simon Fitzmaurice, who developed ALS in his thirties. He was premiering a short film he had made at the Sundance Film Festival when he first noticed foot drop. After a diagnosis of ALS was made, he continued to embrace life. He and his wife had twins, creating a family of five, and they took a trip to Australia. As he lost more and more ability and was reduced to communicating with his eyes on a computer screen, he wrote and directed a film that premiered to acclaim at the Irish Film Festival. A beautiful documentary was made of his journey, called *It's Not Yet Dark*.

Steve Saling lives in the Leonard Florence Home near Boston. He, too, acquired ALS at a young age. An architect, he designed a special facility at the home that allows him to be independent, controlling everything by the movement of his eyes.

Steve Gleason played safety for the New Orleans Saints of the National Football League. He is known for his blocked punt in a 2006 game, which became a symbol of recovery in New Orleans in the team's first home game after Hurricane Katrina. A statue commemorating the blocked punt is in front of the stadium. In 2011, Gleason announced that he had ALS; his life with the disease was featured in the 2016 documentary *Gleason*. Rather than succumb to the disease, he elected to have artificial ventilation. In 2020, he was awarded the Congressional Gold Medal for his contributions to ALS awareness.

There are countless stories that deserve to be told—of those fighting to find a cure for the disease and of those with the disease who fight as best they can until a cure is found. All of them testify to the human spirit and courage.

CHAPTER FIVE

Parkinson's Disease

In 1817, a general practitioner in London first described cases of what he termed the shaking palsy. His account was based on observing the symptoms and signs of six men between the age of fifty and sixty-five. He wrote, "Involuntary tremulousness motion, would lessen muscular power, in parts not in action and even when supported; with the propensity to bend the trunk forwards and to pass from a walking to a running pace: the senses and intellects being uninjured." James Parkinson was describing the disease that would later take his name. In his detailed, sixty-six-page monograph titled "An Essay on the Shaking Palsy," Parkinson defined what he felt was a new "medical species" that had not yet been classified. He said the syndrome had an insidious onset and progressive, disabling course. Parkinson's monograph was largely accurate, though his statement that "senses and intellects were uninjured" are now known to be inaccurate.

A little more than half a century after Parkinson's monograph, Jean-Martin Charcot focused his attention on the condition Parkinson had described. A gifted painter who spoke four languages, Charcot was known for gathering extensive clinical data and correlating them with anatomical findings during autopsy. Charcot had previously described both MS and ALS. Observing his own patients at the Salpêtrière Hospital in Paris, Charcot identified the shaking palsy's four common symptoms: tremor, rigidity, bradykinesia (slowness of movement), and postural instability. He added two more issues Parkinson had overlooked: micrographia, or small handwriting, and a decrease in facial expressiveness (masklike facies), also called hypomimia. In 1872, Charcot presented a case of the shaking palsy to doctors at Salpêtrière. Recognizing Parkinson's seminal contributions, he suggested naming the disease for the London doctor.

Although Parkinson was the first to describe the disease as a neurological syndrome, Parkinson's has been known throughout history, with descriptions in ancient Egypt, China, and perhaps even in the Bible. Leonardo da Vinci wrote in an anatomical manuscript of those who "move their trembling parts, such as their heads or hands, without permission of the soul; which soul with all its power cannot prevent these limbs from trembling." Many well-known figures have been afflicted with Parkinson's disease, such as actor Michael J. Fox, comedian Robin Williams, and US attorney general Janet Reno. The boxer Muhammad Ali was diagnosed as having Parkinson's, and some even believe that Hitler suffered from a form of Parkinson's disease.

PARKINSON'S SYMPTOMS

In 2003, Martin Samuels, chairman of the Neurology Department at Brigham and Women's Hospital, was in the hospital parking garage when he saw Thomas Graboys, a world-renowned cardiologist, walk to his car. Without thinking about it, Samuels asked Graboys who was taking care of his Parkinson's disease. Although Tom had had symptoms for a few years, he had kept them secret. Samuels, a master clinician, could tell in an instant by the way Graboys walked that he had Parkinson's disease. Although Parkinson's is primarily recognized by its motor symptoms, including tremor, slow gait, and stooped posture—the symptoms that enabled Samuels to diagnose Graboys just by watching him in the parking lot—we now know much more about the disease. Indeed, Dr. Graboys soon developed symptoms of dementia, and he wrote poignantly about the course and impact of his tragic illness, referred to as dementia with Lewy bodies, in his book *Life in the Balance*.

In the book, Graboys describes what happened while he slept. He writes of vivid dreams in which he is playing football or, in a nightmare, punching someone, only to wake up to find that he had been hitting his wife or knocking over the lamp near his bed. We now know that one of the early symptoms of Parkinson's disease is called rapid eye movement sleep behavior disorder. When we sleep normally, we go through a period of REM (rapid eye movement) sleep, and it is during this time that we often have vivid dreams. Under normal conditions, when we have these vivid dreams our muscles are essentially inactive, so we experience the dream but do not move. In some people with early features of Parkinson's disease, this natural inactivity is no longer there. Some dreams can be physically acted out with aggression, sometimes causing injury to a person sleeping next to the patient with early Parkinson's disease.

Interestingly, these sleep disturbances can occur years before the motor symptoms of Parkinson's disease are evident. However, the rapid eye movement sleep behavior disorder is not the only sleep disorder in Parkinson's disease. They may have excessive daytime sleepiness or will vocalize, including cussing, or they may injure themselves with violent movements that can result in fractures. Those with such symptoms have been known to sleep on a floor in a room without furniture so they will not hurt themselves.

Another symptom that may occur before a diagnosis of Parkinson's disease is chronic constipation. As many as half the people suffering from Parkinson's disease have constipation, which may precede the motor syndrome. Studies have shown that people who have fewer than one bowel movement per day have a somewhat greater likelihood of developing Parkinson's disease.

Another non-motor symptom associated with Parkinson's disease that can predate motor symptoms is a decrease in the sensitivity of smell.

Along with constipation and sleep disturbances, difficulty with smell is well recognized in patients with Parkinson's, although this can also occur in Alzheimer's disease and other neurological disorders. In the longitudinal Honolulu-Asia aging study, people were screened for problems with smell by administering an odor identification test. Researchers found impaired smell can be present before clinical Parkinson's disease by several years and could be a useful screening tool to detect those who are at high risk for developing Parkinson's disease later in life. It is remarkable to think that constipation, flailing around in bed while asleep, and trouble smelling could all be early signs of Parkinson's disease.

In addition to these non-motor, prodromal Parkinson symptoms, once a clinical diagnosis is made based on resting tremor, rigidity, and/or slow movement, it becomes apparent that Parkinson's disease may include other features, including a sizeable drop in blood pressure when one stands up, hallucinations, depression, apathy, and mild cognitive impairment that ultimately leads to dementia.

THE CRIME SCENE

One of the major clues at the crime scene was discovered in 1912 by the neurologist Frederic Lewy, who was working with Alois Alzheimer in Germany. Lewy performed a new staining technique on the brains of patients who had had Parkinson's disease and observed small, round abnormalities under the microscope. These abnormalities were inside neurons and were very distinctive protein aggregates (made when proteins accumulate and clump together), later named for him. These Lewy bodies would turn out to hold a major clue for the cause and mechanism of Parkinson's disease.

Another major feature of the crime scene is death of nerve cells in a region of the brain called the substantia nigra, which means "black substance" in Latin. If one looks at the brain stem, the normally darkly pigmented neurons of the substantia nigra are not seen in patients with Parkinson's disease. In 1919, a neuropathologist named Constantine Tretiakoff

PARKINSON'S DISEASE

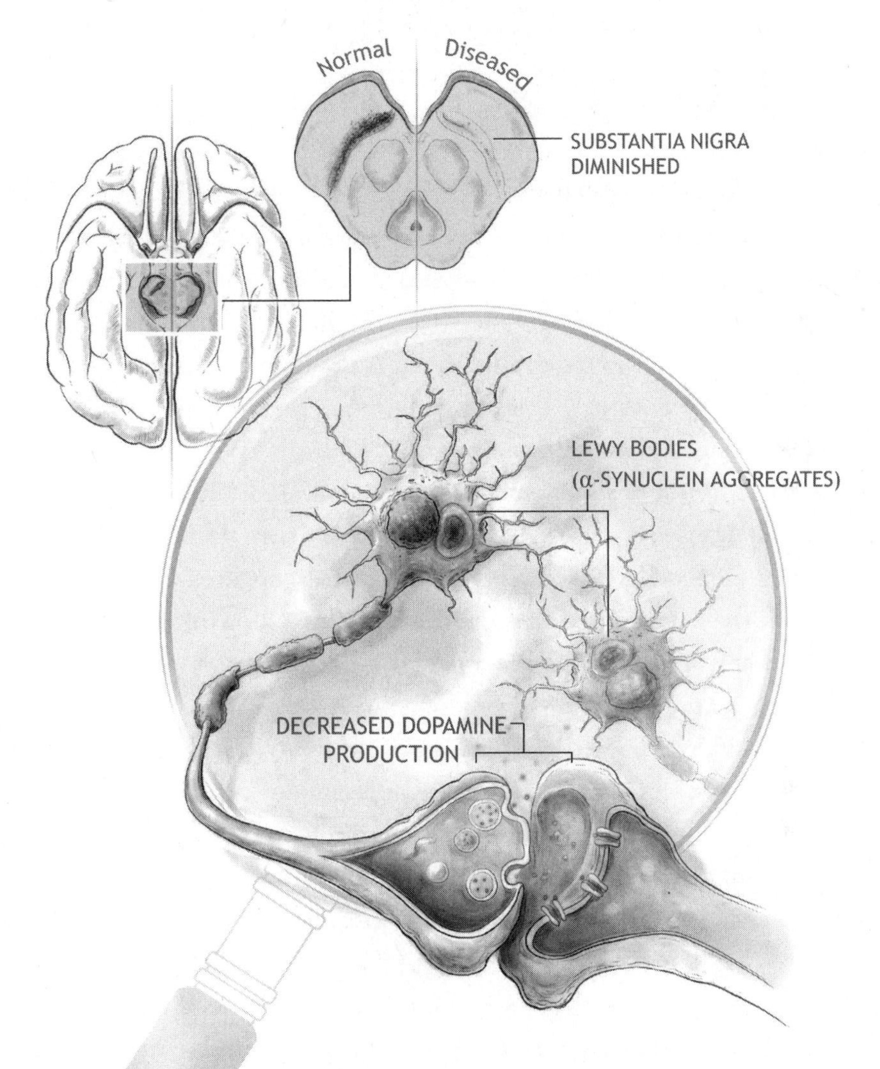

The Parkinson crime scene shows nerve cells lost from a normally darkly pigmented region in the brain stem called the substantia nigra, dopamine neuronal loss, and deposits of α-synuclein aggregate that form Lewy bodies.

tried to make a connection between the loss of nerve cells from the substantia nigra and Parkinson's disease. He observed changes in the substantia nigra both in Parkinson's disease and in a special type of parkinsonism that comes on after a viral encephalitis, a condition called post-encephalitic Parkinson's disease.

A major breakthrough in understanding the crime scene came decades later when Oleh Hornykiewicz and Herbert Ehringer at the University of Vienna measured a substance called dopamine in the brains of people who had had Parkinson's disease. Until the late 1950s, dopamine was thought to be a minor player in the synthesis of noradrenaline, a hormone and neurotransmitter. However, a series of experiments caused researchers to wonder if dopamine did something more. In 1957, an article in *Nature* described how the dopamine precursor, L-DOPA, could reverse a Parkinson-like effect in animals caused by a drug called reserpine. Two years later, researchers found comparable amounts of dopamine and noradrenaline in certain nerves in the brain, suggesting dopamine might have a distinct regulatory function of its own. Another bit of evidence: dopamine lowered the blood pressure of guinea pigs, while noradrenaline raised it.

What truly intrigued Hornykiewicz was a 1959 study by Åke Bertler and Evald Rosengren at the University of Lund, Sweden, showing that the highest concentration of dopamine in the brain occurred in an area called the striatum, which has functions that include facilitating voluntary movement. Later, Hornykiewicz wrote about reading Bertler and Rosengren's study, saying, "It was as if something like scales had fallen from my eyes." He didn't have to read the study twice. "In a flash, I saw in my mind the brain dopamine riddle solved." Hornykiewicz and Ehringer began to collect fresh brains at autopsy from patients who had died with Parkinson's and age-matched, unaffected people. Some of his senior colleagues told him not to waste his time.

Hornykiewicz and Ehringer began by analyzing several brains from patients without Parkinson's disease and indeed detected dopamine in the substantia nigra. The first time they analyzed a patient with Parkinson's they made a dramatic discovery. Dopamine is measured by an iodine color reaction—when an iodine solution is applied to the tissue, the brain samples turn pink if dopamine is present. In the samples from Parkinson's patients, Hornykiewicz immediately saw there was no pink, confirming that dopamine was dramatically decreased in the substantia nigra of patients with Parkinson's disease.

I have had such exhilarating experiences in the lab. We study molecules, proteins, and genes that can't be seen either with the naked eye or even under the microscope. Their presence is shown via a chemical reaction that creates a color or a line on a gel. I remember the times I stood by a machine and raised my hands in the air in excitement when I saw a color develop or a gel blot appear, indicating that an experiment had worked. However, before getting too excited, it is important to see the finding in multiple samples. Then it has to be repeated. I have also experienced the big disappointment when an apparently successful experiment can't be replicated.

Oleh Hornykiewicz, with his naked eye, had seen how the brains from people with Parkinson's had lowered dopamine levels in the substantia nigra. Over twelve months, Hornykiewicz and Ehringer analyzed the brains of twenty different patients, six with Parkinson's disease and fourteen others who were either normal or had other neurological diseases. Only the brains of the Parkinson's patients had a decrease in dopamine in the striatum. It was not a large conceptual leap to ask the next question: If dopamine was not found in the substantia nigra of Parkinson's patients, what would happen if you gave those patients dopamine? Would it reverse their symptoms?

The compound used was called L-DOPA, a chemical precursor of dopamine. Hornykiewicz gave his supply of this chemical to neurologists at the largest municipal home for the aged in Vienna, encouraging the doctors to give it intravenously. Initially, they were reluctant, but after six months of his pestering, they finally agreed. The effect of the first L-DOPA injections was nothing short of spectacular. Bedridden patients who were unable to sit up could now stand, and those who could not walk suddenly walked around normally.

Oliver Sacks, a neurologist in New York, read about these findings. He was taking care of people with a form of parkinsonism that had developed in individuals following a viral encephalitis outbreak in 1919. The patients on his ward appeared frozen, and when Sacks gave his patients L-DOPA, he, too, observed dramatic reversals of symptoms. Sacks then wrote a book about his experience, called *Awakenings.* It was made into a movie, with Robin Williams playing Oliver Sacks and Robert De Niro playing one of the patients. Occasionally, there are dramatic moments in medicine—this was one of them. Many patients with disease dream of an injection that will relieve their symptoms or a pill that will permanently take away their pain. I often say to my patients with MS or Alzheimer's that I wish I could pull something out of the drawer and give it to them, but the chances for that are slim. Oliver Sacks was lucky enough to have done it.

A major part of the Parkinson crime scene was now identified, a fingerprint indicating that thieves had stolen dopamine in the substantia nigra and striatum. However, nothing is as simple as it seems. As Sacks and Hornykiewicz learned, L-DOPA's dramatic effects didn't last forever. A major problem in treating Parkinson's patients with dopamine is an increase in dyskinesia, a particular type of abnormal movements. Something similar occurred with insulin and the treatment of diabetes. Before insulin was synthesized and used to treat diabetes, people did not survive the disease. With the advent of insulin treatment, dramatic healing took place. Many people with diabetes now take insulin and enjoy relatively healthy lives. Nonetheless, many of the complications of diabetes persist although there is metabolic control provided by insulin.

Not everyone believed Oleh Hornykiewicz's findings. Some who could not repeat the findings most likely chose the wrong patients or gave an incorrect dose. In medicine there are always naysayers—something I experienced with my early treatment of MS with cyclophosphamide—and a landmark paper published in the *New England Journal of Medicine* in 1967 silenced Hornykiewicz's skeptics. A team led by George Cotzias in Long Island tested two compounds in patients with Parkinson's: one was L-DOPA, the other was melatonin-stimulating hormone. Cotzias chose melatonin because this hormone had been reported to be present in the basal ganglia but was apparently decreased in Parkinson's. As it turned out, high doses of L-DOPA made the Parkinson's patients dramatically better, but melatonin made patients temporarily worse. It is helpful in an experiment to have something that makes the biology worse and something else that makes it better, because it allows one to zero in on the relevant biologic pathways. Cotzias reported that the improvement he saw when high doses of L-DOPA were given occurred within two to three hours.

The crime scene in Parkinson's was beginning to yield to careful investigation, and the pathway for treatment with L-DOPA was established. However, there was another critical aspect to the crime scene that had not been identified: what was the nature of the neuronal Lewy bodies that were observed in the substantia nigra and elsewhere in the brains of patients with the disease? Although Lewy bodies were seen under the microscope in 1912 and named after the person who discovered them, what was happening with this seminal part of the crime scene was not identified until 1997.

The discovery was published in two successive papers. One paper, by Mihales Polymeropoulos and colleagues at the NIH, reported on the genetic analysis of an Italian family, known as the "Contursi kindred," who suffered from a rare familial form of Parkinson's. Based on this family, Polymeropoulos et al. were able to identify the causative Parkinson's gene as one that coded for the neuronal protein alpha-synuclein. Shortly afterwards, in a second paper, Maria Grazia Spillantini and coworkers at Cambridge University, England, used this genetic discovery to show that Lewy bodies in postmortem brain sections of Parkinson's patients contained alpha-synuclein. The key component of the crime scene was thus discovered.

The first part of the discovery of alpha-synuclein took a path based on genetics. Although there is no increase in the incidence of Parkinson's in identical twins versus fraternal twins, Lawrence Golbe and Robert Duvoisin identified a group of Parkinson's patients from New Jersey who appeared to come from the same family centered around Contursi, Italy. In 1990, they reported on a dominant form of Parkinson's disease that affected forty-one individuals in two families, thirty-one from one family and ten from another. Both families, or kindreds, had emigrated to New Jersey and New York between 1890 and 1920 from Contursi, a village in the hills of the Salerno province in southern Italy. The investigators identified sixty individuals in five generations known to have typical Parkinson's disease. This finding is similar to what was seen in Alzheimer's disease in the Colombian family with a dominant mutation in the presenilin gene. As discussed in the chapter on Alzheimer's disease, the presenilin gene provided an important clue to the role of amyloid protein in Alzheimer's, as it is the enzyme that creates the toxic beta-amyloid protein.

Thus, the possibility existed that genetic analysis of the Contursi family could unlock a secret to the cause of Parkinson's disease. Robert Nussbaum's team at the NIH did sophisticated blood analysis of the Contursi family, performing analysis similar to what Bob Brown did in identifying the *SOD* gene in ALS. They pinpointed the Contursi gene mutation to a small region of chromosome 4. They then sequenced what they thought was the abnormal gene and put it into the NIH gene bank, to which all scientists had access. The gene they identified was called *SNCA*. The gene coded for a protein called alpha-synuclein. The *SNCA* gene sequence had also been placed in the gene bank by two other scientists.

Finding a disease-causing gene is a unique event in science. Polymeropoulos and colleagues' paper was published in the journal *Science* on June 27, 1997, only a month after it was submitted, very fast for the publication of a paper in *Science*. In the modern era, it typically takes a long time to publish a paper in a high-level journal such as *Nature* or *Science*. Months or even years are required to answer reviewers' comments and to perform additional experiments requested by the reviewers.

The second part of the discovery of alpha-synuclein in Parkinson's disease was communicated in a two-page paper submitted in 1997 to *Science* as scientific correspondence. It contained only two figures. Having found a mutation in the alpha-synuclein gene in the Contursi family, Maria Grazia Spillantini wanted to see if antibodies that were specific for alpha-synuclein could stain tissue from patients with both Parkinson's disease and Lewy body dementia. She used an antibody that she had in the freezer that was previously made to study the alpha-synuclein protein. She took tissue that had been sitting in the lab from patients with Parkinson's disease and added the alpha-synuclein antibody from the freezer. Looking under the microscope, she easily saw that the Lewy bodies of Parkinson's disease were stained by her antibodies. This experiment elicited results that could not have been more clear. The mysterious Lewy bodies of Parkinson's disease were identified as being composed of alpha-synuclein.

Other genetic studies followed. Singleton and his group from the Laboratory of Neurogenetics at the NIH identified another family group of dominant Parkinson's disease, named the Iowa kindred. It turned out that the genetic process in these patients did not involve a mutation in the alpha-synuclein gene but rather a duplication of the entire alpha-synuclein gene. These findings suggested that a genetic abnormality in the alpha-synuclein gene that caused Parkinson's disease could not only be related to a small mutation in the gene but also to *too much* of the gene.

This resembles the situation in Down's syndrome in which there is an extra chromosome 21. The APP gene that is linked to Alzheimer's disease is on chromosome 21, thus people with Down's syndrome have an increased incidence of Alzheimer's disease because they have *too much* of the gene. As with Alzheimer's disease and ALS, additional genes have been identified as being related to Parkinson's disease, but with these two classic papers in 1997, alpha-synuclein became the major culprit at the crime scene of Parkinson's disease. The question then remained: Could the alpha-synuclein discovery provide a clue to Parkinson's disease that provided a road map toward a cure?

INVESTIGATING ALPHA-SYNUCLEIN

Given that alpha-synuclein accumulation appears to be at the heart of the major forms of Parkinson's disease, a major line of inquiry is, of course, How does it get started, and can it be averted? Heiko Braak and his colleagues Chris Hawkes and Kelly Del Tredici postulated what they called "a dual-hit hypothesis" for Parkinson's disease. During sporadic Parkinson's long prodromal period, non-motor features develop, such as problems with smell, sleep, and the GI tract. Early sites of such Lewy body appearance are in the olfactory bulb and the neurons of the intestinal system. To connect the dots between the non-motor problems that precede a Parkinson's diagnosis and the Lewy body pathology, Braak and his colleagues speculated that a neurotropic pathogen—probably a virus—enters the brain via two routes. First, the pathogen enters the body through the nose and then progresses into the brain. Second, the pathogen enters the GI tract, perhaps from swallowing nasal secretions in the saliva. This pathogen could then enter the neurons in the intestine and be transferred up a nerve, called the vagus, into the brain, and then to the neurons of the substantia nigra, where the typical motor symptoms begin.

This dual-hit hypothesis suggesting a type of infection has never been proven. Indeed, there's no evidence that there's an infectious agent in Parkinson's disease, although spread of the abnormal synuclein deposits from the gut or nasal cavity into certain regions of the brain might happen. It is also possible that environmental factors could contribute to the alpha-synuclein deposits, because pesticides are known to be risk factors for Parkinson's disease and could be an initial event that triggers alpha-synuclein accumulation outside the brain.

Braak proposed six stages of Parkinson's, beginning with olfactory and gut disturbances in stages one and two. In stages three and four, neurons higher up in the midbrain and forebrain are involved in patients who have motor symptoms, and there are sleep and motor disturbances at these stages. Finally, this progresses to stages five and six, when the limbic and neocortical brain regions are involved, causing symptoms of cognitive problems and dementia. The progression of changes in the crime scene thus appears to mimic the clinical stages, and it is only logical that they would: prodromal Parkinson's disease with constipation and trouble sleeping, followed by early-stage Parkinson's disease, which includes tremor, rigidity, and slowness of movement, followed by cognitive impairment and problems with blood pressure, and finally later-stage disease, including dementia and psychotic symptoms.

In the absence of evidence for an infectious agent in Parkinson's disease, the spreading of the disease described by Braak, in addition to other findings, has raised the question of whether Parkinson's disease has features resembling a prion disorder. Prion diseases are unusual infectious diseases caused by abnormal folding, spread, and accumulation of an otherwise normal protein that has been named the "prion" protein.

Nucleic acids are present in all conventional infectious agents, including bacteria and viruses. Surprisingly, accumulation of the misfolded prion protein can cause rare brain infections in the absence of nucleic acid, DNA, or RNA; their absence is what makes the infectious prion protein unique. Prion diseases include Creutzfeldt-Jakob disease in humans and bovine spongiform encephalopathy, also called mad cow disease, in cattle.

Kuru is another prion disease, a fatal infectious brain disease discovered in Papua New Guinea in the 1950s. Kuru was acquired and spread during funeral rituals in which cannibals either ate infected brain tissue or smeared it on their bodies. When prion diseases like kuru were closely studied, transmission was seen to occur by injecting brain tissue from people with kuru into other primates, and later into mice, each of which became infected and died. In other forms of prion disease, humans can acquire progressive neurological disease by consuming contaminated beef, growth hormone injections isolated from cadavers, blood transfusion, corneal transplants, and implants of incompletely sterilized electrodes into the brain. Researchers have tried to transmit Alzheimer's or Parkinson's by injecting tissue from these diseases into the brains of animals. This would prove that they are prion diseases, but the experiments have been unsuccessful. Nonetheless, some researchers are investigating whether an abnormally folded protein plays a major role in Parkinson's by a prionlike spread throughout the brain. Not everyone believes this is the case. As with every disease, the ultimate proof will need to come from therapies that selectively target the physical movement of protein aggregates and inhibit the progression of the disease.

A study published in 2020 by Per Borghammer in Denmark suggests that there may be two types of Parkinson's disease, one that begins in the brain (brain-first) and one that begins in the body (body-first). In the body-first type, the initial deposits of alpha-synuclein occur in the intestine and then migrate via the vagus nerve into the brain. In the brain-first type, the initial deposits of alpha-synuclein occur in the brain and then move to the intestine. He discovered this by imaging the gut and the brain in patients who did or did not have sleep disturbances. Understanding how the disease starts may ultimately lead to discovering which disease-modifying therapies are most likely to be effective when treatment is given, or to prevent disease in its early stages.

FETAL BRAIN TRANSPLANTATION

Medications that targeted dopamine deficiency helped mitigate the motor symptoms for Parkinson's patients, but for many the disease became debilitating. A dramatic approach for those more severely affected was to transplant human fetal tissue grafts directly into the brain. The logic was that normal neurons in pieces of fetal tissue could secrete dopamine and directly relieve symptoms. It was a bold strategy. Patients often ask me whether one can transplant either tissue or stem cells into the brain to relieve symptoms of MS, Alzheimer's, and ALS. Unfortunately, it is not possible, as the cause of the symptoms involves complex dysfunction of the nervous system. For Parkinson's, it isn't necessary to repair the complex damage in the nervous system, only to provide a source of dopamine.

Initial results of fetal tissue transplants in advanced Parkinson's patients showed improvement in symptoms and perhaps a slowing of Parkinson's disease progression, which understandably created a great deal of excitement. To establish that the observed benefit was real, double-blind, controlled studies were performed. In such a controlled trial, sham surgery was required, which involved patients being anesthetized, having a stereotactic frame attached to their head, and having a partial hole drilled into the skull—all without receiving the transplant. Imagine the difficulty in doing these controlled trials. Unfortunately, despite initial excitement about fetal brain transplants in Parkinson's disease, these sham-controlled studies did not show clinical benefit. The disappointing results came as a big surprise to all involved, as there appeared to be dramatic improvement in some of the first people who received the transplants outside the bounds of a controlled study.

Scientists examined the brains of those who died within eighteen months after the transplant, to determine what had happened to the transplanted neurons. They found that the neurons survived and appeared unaffected by Parkinson's disease. However, because we know that Parkinson's disease takes time to develop, this wasn't definitive proof that the transplant was successful. The long-term effects were discovered when investigators examined the brain of a woman who had received a transplant fourteen years earlier—the longest survival of a person with a transplant at the time—and found something dramatic.

The findings were published in 2008 in the journal *Nature Medicine.* They reported on a woman who was diagnosed with Parkinson's in her late thirties and experienced motor problems that weren't controlled by medications. She lived with Parkinson's for twenty-two years before undergoing a fetal tissue transplant in 1993. The transplant consisted of brain neurons isolated from four embryos obtained one to two months after conception.

The neurons were isolated from the embryos and injected into the woman's substantia nigra, the area damaged in Parkinson's disease.

After transplantation, she experienced improvement in her "off time," that is, the amount of time that she did not require medication or have abnormal movements. She also required lower amounts of L-DOPA. However, ten years after transplantation, her Parkinson's disease worsened. She had difficulty with gait and balance and began falling, which could not be alleviated by medication. She died fourteen years after the transplantation.

Her treatment provided a unique opportunity to examine the brain and determine what happened to the graft. Although the graft survived, scientists were struck by what they saw in the implanted fetal tissue: some grafted neurons contained aggregates of alpha-synuclein. Many of these looked like typical Lewy bodies. In other words, the normal implanted brain tissue took on the appearance of Parkinson's disease. Amazingly, it appeared that the disease had spread from neurons in the woman's brain into the implanted fetal neurons. This had never been seen in previous cases of fetal brain transplantation, though investigators had not previously had the opportunity to evaluate someone fourteen years after grafting. Remember that prodromal non-motor symptoms of Parkinson's such as constipation, sleep disorders, and smelling dysfunction, can occur as much as twenty years before the motor symptoms. Parkinson's is classified by what occurs at the crime scene—the deposition of Lewy bodies and the presence of alpha-synuclein. The authors concluded that the alpha-synuclein aggregates of Parkinson's disease were capable of spreading to healthy neurons and making them Parkinson-like. Thus, Parkinson's appeared to have properties of a prionlike disease.

A process in which normal transplanted tissue is altered by the host environment has been observed in type 1 diabetes, although in this disease it is a property of the immune system. Type 1 diabetes, like multiple sclerosis, is an autoimmune disease in which the immune system attacks one's own tissue. In MS, the immune system attacks the brain. In type 1 diabetes, it attacks islet cells in the pancreas. Islet cells make insulin, and with the immune destruction of islet cells and loss of insulin, the symptoms of diabetes arise. Children who develop clinical diabetes may have 80 percent of their islet cells already destroyed before they have clinical symptoms, just as someone who develops Parkinson's disease may have had neuronal loss going on many years before the disease manifests itself clinically. In experiments in which healthy islet cells from an identical twin were transplanted into the twin with diabetes, the abnormal immune system of the twin destroyed the healthy transplanted cells.

In Parkinson's, the abnormal alpha-synuclein itself affected the transplanted tissue. Is there experimental evidence to show that Parkinson's may be a prionlike disease and that alpha-synuclein can spread it? To evaluate the causative relationship between a microbe and a disease, we use Koch's postulates, which are four criteria for determining cause of disease, published by Robert Koch in 1890. Koch's postulates state that if one has identified a disease-causing organism or process, one should be able to transfer it to a person or animal and show that it causes the disease. This, of course, is the basis of identifying viruses that cause disease, such as HIV. In MS, a similar conceptual approach has shown that T cells can transfer disease between animals. When brain-reactive T cells are transferred into animals, they cause an MS-like disease. The muscle disease myasthenia gravis is caused by antibodies directed against a receptor that is important for nerve-muscle communication. It is possible to transfer myasthenia gravis from people into experimental animals by injecting antibodies from people with the disease into animals. It is possible to transfer Parkinson's between cells in tissue culture. When investigators placed altered synuclein proteins in tissue culture, the altered synuclein proteins moved from one cell to another. A direct in vivo experiment demonstrated that injecting Lewy body extracts taken from the brains of Parkinson's patients into the brains of mice or monkeys triggers alpha-synuclein pathology similar to what is seen in Parkinson's disease.

There is a body of evidence that suggests Parkinson's may be an atypical prionlike disorder driven by alpha-synuclein. I discussed this concept in detail with Dennis Selkoe, who agrees on the importance of alpha-synuclein in Parkinson's disease but believes there is not convincing evidence that Parkinson's is a prionlike disease. We know that alpha-synuclein can affect the nervous system in different ways. Another disorder of alpha-synuclein is called multiple-system atrophy, which has a widespread constellation of neurological signs and symptoms that differ from Parkinson's disease.

INSIGHTS INTO PARKINSON'S FROM AN UNEXPECTED SOURCE

Before exploring alpha-synuclein, let's examine a dramatic, serendipitous event in the history of Parkinson's research. It occurred in Northern California when addicts took a drug sold as a synthetic heroin and came down with a disease that looked like advanced Parkinson's disease. J. William Langston, then director of neurology at the Santa Clara Valley Medical Center in San Jose, was the lead investigator in this medical detective story.

It began when he was called to see a patient in the locked psychiatry unit who was alert but had no spontaneous movement. Although the patient looked like an advanced Parkinson's patient, the man was in his early forties, and his symptoms had come on overnight. Langston and his colleagues then found six other similar cases in the San Jose area. All of them had taken a street drug contaminated with the compound N-methyl-4-phenyl-1, 2, 3, 6-tetrahydropyridine, known by the acronym MPTP. The addicts who developed this Parkinson's-like syndrome dramatically recovered when they were treated with dopamine, demonstrating that an environmental toxin could trigger Parkinson's disease. However, MPTP did not re-create Parkinson's classic alpha-synuclein deposits and was simply toxic for dopamine neurons. In the study of Parkinson's disease, toxin-based models based on killing dopaminergic neurons have proven useful in developing treatments for dopamine replacement therapy, but they have not yet identified agents that can slow, halt, or reverse the disease course because they do not target the alpha-synuclein imbalance in Parkinson's.

The fascinating MPTP story turned out to be a sidebar, and investigators have turned their attention to alpha-synuclein as the basis for developing disease-modifying Parkinson's therapies. Among the possibilities: facilitate the clearance of misfolded alpha-synuclein or block the neuron-to-neuron passage of alpha-synuclein through the brain.

HOW ALPHA-SYNUCLEIN WORKS

Although we don't fully know the normal function of alpha-synuclein, we believe that it helps regulate the movement of vesicles in neurons, tiny sacs that transport molecules in the cell and are perhaps involved in dopamine release. Alpha-synuclein has also been shown to interact with lipids and thus influence cell membrane structure. Because it is also found in white blood cells, red blood cells, and platelets, it may function outside the nervous system in immune system development.

In its nontoxic form, alpha-synuclein exists as a single structure, like a straw, called a monomer. However, when several straws clump together, they are called oligomers, and when the clumps aggregate and twist upon each other, they become toxic and are called fibrils. The aggregated-clumped fibril is caused by a misfolding of the alpha-synuclein protein, which leads to the formation of Lewy bodies. Thus, Parkinson's is considered a protein misfolding disease. Misfolded alpha-synuclein increases stress on the cell and can impair its ability to maintain the proper folding of individual proteins. Alpha-synuclein also promotes inflammation, which includes activation of microglia.

Different biologic processes cause disease. In cancer, cells grow uncontrollably. In autoimmune diseases like MS, the immune system attacks its own tissue. And in Parkinson's, Alzheimer's, and prion diseases like Creutzfeldt-Jakob disease, when a protein folds in the wrong way, it can damage neurons and cause a progressive neurological disease.

When we consider alpha-synuclein-based animal models of Parkinson's disease, these models may be imperfect, but they are crucial for understanding the disease and developing drugs to treat it. There are successful treatments for human disease that were not first tested in animal models, but this is generally not the case. Animal models paved the way for current MS drugs, and, as we discussed, Alzheimer's animal models based on A-beta have provided the impetus for development of A-beta therapeutics. At scientific conferences it is common to hear that animal models are not the actual disease, but in virtually all my discussions with pharmaceutical companies about developing new therapies, the question invariably arises of how does a candidate drug perform in the animal model. It is clear that rodent models of alpha-synuclein abnormalities are likely to pave the way for disease-modifying drugs in Parkinson's disease.

What, then, are these animal models? Because there is loss of dopaminergic neurons in specific brain regions, leading to the classic motor symptoms of Parkinson's disease, toxin-based models were initially developed, targeted at killing dopaminergic neurons. This approach took advantage of the discovery that MPTP caused a Parkinson-like disease in drug addicts. Although MPTP models have been used to develop treatments for Parkinson's symptoms, with the discovery of alpha-synuclein in Lewy bodies, it became clear that MPTP models did not represent true Parkinson's disease.

In order to come up with treatments that could modify the disease course and treat patients with Parkinson's disease prior to their developing motor symptoms, the models needed to involve altering alpha-synuclein. One of the primary focuses of Parkinson's disease models and disease-modifying therapy is to create mice that have an excess of human alpha-synuclein, determine if they show signs and symptoms of Parkinson's disease, then develop strategies to lessen or neutralize toxic forms of alpha-synuclein.

Broadly speaking, there are three types of alpha-synuclein animal models. The first are genetic models, in which an abnormal human alpha-synuclein gene is inserted into mice and the mice make alpha-synuclein. This is similar to what we described in both Alzheimer's disease with the A-beta protein, and in ALS with the SOD protein. A second way to study alpha-synuclein is to create a virus that delivers the abnormal alpha-synuclein aggregates to the brain. The third involves transmission models, in which alpha-synuclein itself is injected into the brains of animals.

Investigators have attempted to mimic the prodromal features of Parkinson's disease by injecting alpha-synuclein aggregated fibrils into the brain area that controls smell. They found that the injected material spread to other parts of the brain, so mouse models of prodromal Parkinson's disease can be used to study the progression of the disease. Other attempts to generate models of Parkinson's were done by injecting human alpha-synuclein aggregates into the rat brain in combination with virus-mediated expression of human alpha-synuclein. This approach was found to enhance the ability of aggregated alpha-synuclein to trigger Lewy-like pathology in dopaminergic neurons. Additionally, there was a prominent inflammatory response, including both activation of resident microglia and infiltration of white blood cells.

Perhaps one of the most provocative new animal models was created in our center by Dennis Selkoe and his colleagues, Silke Nuber and Ulf Dettmer. Their work in cultured cells and in mice began in 2011, and it has led to a new way of thinking about alpha-synuclein in Parkinson's disease, as well as potential new treatments. Dennis had long been working on A-beta and its role in Alzheimer's disease. In 2008 he reported in *Nature Medicine* that he could isolate clumps of human A-beta oligomers directly from postmortem Alzheimer brains and inject them into healthy adult rats and make the animals forget a learned task. This is like the studies where investigators injected alpha-synuclein from Parkinson's brains into mice in an attempt to recapitulate some of the features of Parkinson's disease.

Tim Bartels joined Dennis's lab in 2010 and began investigating the normal state of alpha-synuclein in both brain cells and red blood cells, another major site of alpha-synuclein expression. When Bartels and Selkoe purified alpha-synuclein under special conditions using sequential fractionation, they found it had a folded structure, counter to the long-held assumption that it was mostly unfolded. Moreover, they found that it weighed four times more than what others had reported. This led them to propose that alpha-synuclein, in its normal condition, was not a single strand but four identical strands wound around each other in a quadruple protein complex called a tetramer. Bartels found that the four-stranded complex resisted abnormal aggregation, compared to single strands, which did not. If this was true, then the non-diseased form of alpha-synuclein was in the form of a four-stranded complex.

Dennis thought they were able to make this observation when others hadn't because in studying alpha-synuclein, they didn't use the common practice of breaking cells open and dissolving their contents in detergents. Indeed, they showed that if one used detergent, the four-stranded complex was broken into single strands. In a 2011 article in *Nature,* Dennis and his team argued that alpha-synuclein became pathogenic when the four-stranded-complex dissociated—that is, it became single strands, which then aggregated to form the toxic form of alpha-synuclein.

A similar concept had been put forward by Jeff Kelly at the Scripps Research Institute for a different protein called transthyretin, a blood protein that existed normally as a four-stranded complex but could fall apart into excess single strands with self-aggregating properties. Kelly and colleagues found that a systemic amyloid disease related to a misfolded protein was caused by a four-stranded complex that had become unstable and unfolded. Kelly's team then invented compounds to stabilize the complex and slow down the fatal form of amyloidosis that occurs in the heart and peripheral nervous system.

Kelly's work supported Selkoe's controversial new hypothesis about the previously unrecognized normal structure of alpha-synuclein. In further support of Dennis's new hypothesis, a group from Johns Hopkins investigated a neurological condition called Gaucher's disease, which leads to abnormal aggregation of alpha-synuclein. Using Dennis's methods, they showed that this condition also causes an abnormal decrease in the four-stranded alpha-synuclein complex and even identified a drug, called miglustat, that may overcome the abnormality.

If accurate, Dennis's discovery opened up a new way to treat Parkinson's: find a way to stabilize the physiologic four-stranded complex of alpha-synuclein so it doesn't dissociate into single strands. This could reduce alpha-synuclein toxicity and provide the basis for a new treatment for Parkinson's disease. Up to that point, work had been done only in tissue culture. A major question was whether these findings—when translated into a mouse—would recapitulate the features of Parkinson's. Developing disease-modifying treatments requires animal models that faithfully reproduce the cardinal features of the disease.

With Silke Nuber joining Dennis's team, a genetic mouse was created that expressed a mutant form of alpha-synuclein that could no longer form the normal four-stranded complex. Remarkably, these mice developed classic features of Parkinson's disease. At three to four months of age they had a resting tremor and could no longer walk or climb normally. Even more remarkable, when their brains were examined under the microscope, they had round, Lewy body–like inclusions in dopaminergic and other neurons. I remember touring the laboratory with Ann Romney on the exciting day that a fully formed Lewy body was seen by electron microscopy in a mouse. The mice not only had neuronal changes and behavioral features like Parkinson's disease; they also had degeneration of dopaminergic fibers and neurons. Finally, their abnormal gait and climbing was improved by treating them with L-DOPA.

As often occurs with an unexpected and provocative discovery that proposes a new paradigm, Dennis had a lot of trouble getting grant funding for this new work. He met with resistance from the scientific community. Since then, others have now confirmed his findings. More importantly, the new mouse model of Parkinson's they created provides a unique tool to understand the normal four-stranded form of alpha-synuclein and devise treatments. Following Dennis's publication in *Neuron* in 2018, scientists are anxious to study the new mouse model. Dennis's team has deposited the mice at the commercial Jackson Animal Laboratory in Maine, so they are widely available.

Clemens Scherzer in the Ann Romney Center has spent his career studying Parkinson's disease, and he's used a unique approach to find a treatment. It began with basic studies of cell lines in the test tube and wound up with studying the entire population of Norway. Clemens's idea was to determine whether he could find compounds that would lower synuclein levels in cells. He did a high-throughput screen using robotics that could assay hundreds of wells at a time, searching for compounds that altered alpha-synuclein gene expression in neuronal cells. He performed the screen by culturing neuroblastoma cells and measuring the synuclein gene expression after exposing the cultures to 1,126 different chemical compounds. The compounds included drugs approved by the FDA and a diverse set of natural products, including vitamins, health supplements, and alkaloids. Cells were treated with the compounds for forty-eight hours. He found that thirty-five compounds changed synuclein gene expression levels. He then measured levels of the alpha-synuclein protein to make sure that the changes in the synuclein gene also affected protein expression. Through this process, he discovered that drugs that activate the beta-2 adrenalin receptor can decrease alpha-synuclein gene expression.

Scherzer then tested his discovery both in a stem cell–derived neuronal culture and in mice with neurotoxin-induced Parkinson's disease. He hypothesized that something that blocked the beta-2 adrenalin receptor would increase alpha-synuclein gene expression levels, which in turn could increase alpha-synuclein accumulation and thus increase the risk of Parkinson's disease. On the other hand, he hypothesized that something that stimulated the beta-2 adrenalin receptor would promote dopamine neuron health by reducing alpha-synuclein gene expression and the accumulation of toxic alpha-synuclein.

Scherzer was able to test this hypothesis in people without doing a clinical trial of compounds that targeted the pathway he had identified because they had already been approved for use in other conditions. From his compound screen, Scherzer knew which drugs increased alpha-synuclein levels and which decreased them. To address the question in people, he turned to a database of 4.6 million Norwegians.

The beauty is that in Norway, virtually all medical diagnoses and medication data have been collected. Scherzer asked whether people taking a common asthma medicine called salbutamol, which blocks beta-adrenergic receptors and penetrates the brain, would have a reduced risk of Parkinson's disease. He found that they did. On the other side of the equation, a drug called propranolol, which is used to treat hypertension by stimulating beta-adrenergic receptors, was associated with an increased risk of Parkinson's disease. Although direct causation was not shown, the study offered clues for potential new treatments for Parkinson's disease. Moreover, they showed how targeting the normal expression of a human disease gene could be a useful strategy for investigating treatments for other diseases. Scherzer's strategy can be more generally applied to the rapid discovery and translation of existing therapeutics for brain diseases. The beauty of his study is that it went from basic studies in the test tube to an examination of data from an entire country's population.

Another major approach being explored to decrease alpha-synuclein is one that has been borrowed from Alzheimer's disease—using antibodies. Investigators are using the immune system, specifically antibodies directed against alpha-synuclein, for treating Parkinson's disease. In Alzheimer's disease, antibodies are directed toward the toxic A-beta protein. An important difference between Alzheimer's and Parkinson's disease is that the toxic A-beta protein in Alzheimer's disease is located outside the cell, whereas the toxic alpha-synuclein protein in Parkinson's disease is found inside the cell. Because antibodies can't penetrate into cells, the alpha-synuclein may not be accessible to antibodies. However, recent studies have shown that small quantities of alpha-synuclein do exist outside the cell. In addition, as discussed above, the progressive deposit of alpha-synuclein inside cells could depend on the movement of misfolded alpha-synuclein between cells, where the antibodies could easily interact with the toxic protein. These findings have led companies to initiate clinical trials in which they administer antibodies directed against alpha-synuclein to treat Parkinson's disease.

As with Alzheimer's disease, the most advanced development of immune therapy in Parkinson's disease is using antibody infusions in which antibodies directed against alpha-synuclein are given intravenously. A number of studies have been published in mouse models, all showing clearance of alpha-synuclein with reduction of symptoms and slowing of the disease process in animals treated with antibodies against alpha-synuclein. These antibodies may act by blocking the interaction of toxic alpha-synuclein proteins outside the cell, which could then lead to a decrease of alpha-synuclein inside the cell. We know from the trials of antibodies against A-beta in Alzheimer's that a small amount of antibody given intravenously enters the brain, where it can exert its effect.

The two most advanced antibodies are being developed by Prothena Biosciences and Biogen, the company that performed the large Phase 3 trial of the antibody aducanumab, which is directed against A-beta. Prothena has successfully administered PRXOO2, a monoclonal antibody directed against alpha-synuclein, in healthy volunteers. It was given in increasing doses and showed no toxicity. PASADENA, a Phase 2 study in early Parkinson's disease, was then undertaken in collaboration with Roche. The double-blind, placebo-controlled, three-arm study tested one-year treatment with two active doses and one placebo given intravenously once every four weeks. Although the trial did not meet its primary endpoint—change from baseline in the Unified Parkinson's Disease Rating Scale—secondary measures were positive, including reduced motor decline and delayed time to clinically meaningful worsening of motor progression over one year. Further trials are planned. The Biogen anti-alpha-synuclein antibody did not succeed in a phase 2 trial and Biogen discontinued their program.

Should the treatment of Parkinson's disease with antibodies directed against alpha-synuclein be successful, it would be a major breakthrough in treatment and lead to the development of other immune approaches that could be more easily administered to large segments of the Parkinson's population. The big question is whether or not treating a person with Parkinson's disease once motor symptoms arise is too late. If treating earlier is the answer, this then presents the challenge of being able to identify patients with prodromal Parkinson's disease so they can be treated before the disease progresses.

Non-immune-based approaches are also being taken to target alpha-synuclein. These include decreasing the activity of the alpha-synuclein gene, as we described earlier, by targeting the beta-2 adrenal receptor. This could involve using drugs that are currently employed to treat other diseases, such as asthma. Other strategies include drugs that inhibit alpha-synuclein uptake, such as heparin; drugs that influence alpha-synuclein aggregation; and drugs that stimulate alpha-synuclein degradation. A natural product called squalamine inhibits alpha-synuclein aggregation and suppresses its toxicity. Other compounds have been reported that displace alpha-synuclein from the membrane. Targeting alpha-synuclein in the fight against Parkinson's will involve strategies for clearance, reducing aggregation, and reducing toxicity, or even reducing production by shutting down the alpha-synuclein gene through gene therapy in a manner that has been used successfully to treat children with spinal muscular atrophy. The success of these therapies will depend on the degree of efficacy relative to side effects.

THE MICROBIOME

Although Parkinson's disease is a brain disease, there is strong evidence that it is somehow related to the gut. Constipation may be one of the first symptoms of Parkinson's disease. One theory is that deposits of alpha-synuclein begin in nerve cells in the gut and then travel up to the brain. But there is more evidence that suggests the gut itself may be involved. In this case, it relates to the gut microbiota, the trillions of bacteria in our digestive tract. We found abnormalities of the gut microbiome in MS, and there are reports of it in Alzheimer's disease and ALS. The evidence that something may be going on in the gut may even be more compelling for Parkinson's disease.

A landmark study of the gut microbiome in Parkinson's disease was carried out by Sarkis Mazmanian at Cal Tech, who found that gut microbiota can cause motor defects and brain inflammation in a mouse model of Parkinson's disease. The investigators studied a model of Parkinson's disease in which mutant human alpha-synuclein is genetically expressed in the mouse. Like Parkinson's patients, these mice have defects in motor function and gut motility—digestion.

To test whether the microbiome affected Parkinson's signs in these mice, they put the animals under germ-free conditions—the mice had virtually no bacteria in the gut at all. They measured motor function in the mice by their ability to crawl on a beam and go down a pole, and by their hind limb clasping reflexes. The animals without a gut microbiome did not develop motor dysfunction or constipation. Meanwhile, the mice with normal microbiota had all the hallmarks of Parkinson's disease. They then measured the amount of alpha-synuclein in the brain and found that the germ-free mice had less alpha-synuclein aggregates. These changes were not only seen in mice raised in a germ-free condition but also in mice treated with high-dose antibiotics, which also wipe out the microbiome. All of these mice had fewer Parkinson symptoms and alpha-synuclein deposits. Not only was there less alpha-synuclein in the brain but the microglial immune cells in the brain were less activated when the microbiome was depleted; they were more like Dr. Jekyll. This suggests that the microbiome can affect brain function.

The next task was determining how the microbiome was causing these effects in the brains of these mice. The investigators found lower short-chain fatty acid concentrations in the feces of germ-free and antibiotic-treated animals. This suggests short-chain fatty acids accelerate the alpha-synuclein aggregation. By administering short-chain fatty acids, they were also able to accelerate alpha-synuclein aggregation in the mice.

These discoveries are all well and good, but showing that the condition of the microbiome of animal models is required for the disease does not necessarily translate to people. To address this question, the investigators asked whether human gut microbes can cause disease when transferred into the mouse model. They took the animals that were germ free and protected from the Parkinson's symptoms and asked the following question: Would the microbiome from a Parkinson's patient worsen disease when transferred into the animals? They took fecal samples from six people diagnosed with Parkinson's disease, as well as from six healthy controls. They only chose patients with new-onset Parkinson's who had not yet received treatment. In a dramatic result, they found that when bacteria from Parkinson's patients were transferred into animals via fecal sample, it caused motor dysfunction in the mice and increased the amount of alpha-synuclein in the brain. This result did not happen when bacteria were transferred from normal individuals or were transferred into animals who did not have a propensity toward Parkinson's.

The results of these experiments suggest there is a pathway by which the bacteria in the gut can impact alpha-synuclein and brain microglia cells in Parkinson's disease. Many of the abnormalities related to alpha-synuclein were affected by the microbiome, including the decreased ability of cells to both clear abnormal proteins and destroy damaged cellular components. Intestinal bacteria have also been shown to modulate the clearance of alpha-synuclein, and it is known that the gut microbiome can produce dopamine and its precursors.

Given these findings, are the changes in the gut primary or secondary to the disease process? As with other neurological diseases, studies of the gut are in their infancy. However, it is theoretically possible that the risk of Parkinson's disease might be partially explained by the depletion or loss of protective gut bacteria, a situation that is potentially reversible. After all, there is a direct connection between the gut and the brain via the vagus nerve, which originates in the brain stem and extends down into the belly. Furthermore, stimulating the vagus nerve in the intestine can send signals to the brain, and the vagus nerve could theoretically be a direct conduit by which material from the intestine could pass into the brain. Brain inflammation of cells such as microglia, which appear to play a major role in neurological illnesses, have also been implicated in Parkinson's disease and could be triggered by the gut.

One scenario to explain how the microbiome can trigger Parkinson's disease is that an inflammatory process in the gut causes low-level inflammation and the inflammation creates a condition that increases alpha-synuclein in the gut. In the Pink1 mouse model of Parkinson's, researchers have found that early-life infection with *Citrobacter rodentium* causes inflammation in the gut that leads to Parkinson's-like symptoms when the mice get older. The alpha-synuclein from the gut is then transferred into the brain via the vagus nerve. Interestingly, people who have had the vagus nerve cut for other medical reasons may have a decreased incidence of Parkinson's disease. Alpha-synuclein is present in the healthy human appendix, including forms that are known to accumulate in Lewy bodies. A study that examined 1.6 million people in the Swedish National Patient Registry and the Parkinson's Progression Markers Initiative found that having the appendix removed decades before the age that Parkinson's disease usually develops was associated with a lower risk of disease.

If abnormalities in the gut truly drive Parkinson's disease, then a crucial question is, What causes these abnormalities? We know alpha-synuclein is key to the disease because alpha-synuclein is at the crime scene. The big question is which factors allow the accumulation of alpha-synuclein. Most Parkinson's is sporadic, not familial, and it is not primarily driven by a genetic factor. And although there are environmental factors that are associated with Parkinson's disease, the major risk factor for Parkinson's disease is aging, similar to Alzheimer's disease. As we get older, nerve cells become more vulnerable to damage, potentially leading to Alzheimer's. We also have a less healthy gut, potentially leading to Parkinson's and Alzheimer's.

An increased risk of Parkinson's has been related to exposure to pesticides, consuming dairy products, a history of melanoma, and traumatic brain injury. A reduced risk has been correlated with smoking, caffeine consumption, physical activity, and urate concentrations. Urate is an antioxidant present intracellularly and in all body fluids. Higher urate concentrations can help other neurodegenerative diseases including Alzheimer's, Huntington's disease, and ALS. Smoking is an unexpected protective factor, and a clinical trial of nicotine is being carried out in early Parkinson's disease. Caffeine, too, may have a protective effect, which could be related to its effect on adenosine receptor antagonism.

These factors raise the question of nature versus nurture and the extent to which Parkinson's disease is intrinsic to brain cells, as opposed to being related to extrinsic factors such as the gut or the environment. Vik Khurana, a neurologist at our hospital, was among the first to create pluripotent stem cell models for Parkinson's disease. In order to evaluate potential extrinsic factors, he is studying a unique family in Bilbao, Spain. The family carries a mutation in the alpha-synuclein gene, and each generation is afflicted by severe Parkinson's disease and dementia. However, there are family members who escape disease even though they have the genetic abnormality. Khurana's team established stem cells from family members to analyze their genetic composition and is now investigating the effect of stool samples from family members on mice with Parkinson's disease to see if the family members who don't get the disease have a protective gut. If he is able to identify gut factors that shield family members with the genetic abnormality from getting Parkinson's, his findings would extend beyond a single family to all others with Parkinson's disease.

Many people ask what they can do to prevent or decrease the risk for diseases like Parkinson's. I certainly wouldn't recommend smoking, but of all the things people can do, physical activity has been shown to be helpful not only for Parkinson's disease but for many other diseases as well. The so-called prudent dietary pattern, which is characterized by high intake of fruit, vegetables, and fish, has been reported to confer a reduced risk of Parkinson's disease.

THE IMMUNE SYSTEM AND PARKINSON'S

As discussed, MS begins as an immune disease in which immune cells infiltrate the brain and attack the myelin sheath that covers neurons. Later, nonimmune factors in the brain can drive MS in the progressive stages.

Alzheimer's and ALS are the opposite, with initial damage beginning in the nerve cells in the brain, followed by a secondary immune response that amplifies the disease. The same is true in Parkinson's disease.

There is evidence that damage to the dopamine system in the brain can lead to an immune response. After damage, there is activation of microglial cells, infiltration of T cells, and changes in astrocytes. The immune system could participate in a self-perpetuating cycle of damage. At the crime scene in people who have died of Parkinson's disease, there's clear evidence of inflammation, with activated microglia and the release of inflammatory factors such as gamma interferon. Even during life, there are signs of inflammation in Parkinson's patients, as PET imaging clearly shows microglial activation.

T cells play a central role in the immune system and are the primary drivers of MS. There is also evidence that T cells may play a role in neurological diseases that are not primarily influenced by the immune system, including Parkinson's disease, Alzheimer's disease, and ALS. Thus, in addition to studying T cells in MS, we have studied T cells in patients with Alzheimer's disease and found that T cells react with A-beta. Regulatory T cells are important in MS, and, as we discussed, regulatory T cells are being tested as a form of treatment in patients with ALS. Indeed, we have found in the ALS mouse model that increasing regulatory T cells can benefit the disease. In Parkinson's disease, recent studies have found that T cells can be isolated that react with alpha-synuclein; T cells infiltrate the brain in mouse models of Parkinson's disease; and regulatory T cells can attenuate disease in some Parkinsonian models. Inflammatory T cells can be detrimental, as was seen when inflammatory T cells were created after investigators treated Alzheimer's disease by immunizing with A-beta and induced encephalitis. Treatment immunization programs in Parkinson's disease are trying to avoid this problem by not immunizing with the parts of alpha-synuclein that trigger inflammatory T cells.

A series of experiments that combined genetics and inflammatory bowel disease provided an important clue to the potential role of inflammation in

Parkinson's disease. The first piece of the puzzle came from genetics. In a major finding reported in 2004, investigators showed that mutations in the *LRRK2* gene was associated with Parkinson's disease; it has become one of the major genes being studied in relation to the disease.

The LRRK mutation is associated with a familial form of Parkinson's disease. Mutations in *LRRK2* are commonly known causes of genetically dominant Parkinson's disease, accounting for approximately 1 percent of sporadic and 5 percent of familial Parkinson's. Some reports suggest that LRRK2 mutations account for up to 13 percent of familial Parkinson's disease associated with dominant inheritance. LRRK2 parkinsonism resembles typical Parkinson's disease with dopamine neuronal loss and deposits of alpha-synuclein. LRRK2 is a molecular switch, turning other proteins in the cell on and off, which can affect neuronal function. The *LRRK2* gene not only affects the machinery of a cell; it also affects the immune system.

The second piece of the puzzle is that Crohn's disease, a form of inflammatory bowel disease, was also found to have functional changes in the *LRRK2* gene. This suggested a genetic connection between inflammatory bowel disease and Parkinson's disease, which created the opportunity for a fascinating experiment. Patients with inflammatory bowel disease are treated with a drug called antitumor necrosis factor, or anti-TNF, a drug that was unsuccessful in treating MS. This led investigators to ask the following question: If the *LRRK2* gene is a risk factor for both inflammatory bowel disease and Parkinson's disease, and inflammation could play a role in Parkinson's disease, is there a decreased incidence of Parkinson's disease in people with inflammatory bowel disease who are treated with anti-TNF drugs? To find the answer, they looked at large epidemiology cohorts.

Investigators considered insurance claims for 170 million people to see if there were any links between inflammatory bowel disease and Parkinson's disease. They found that 28 percent of the people with inflammatory bowel disease were at risk for Parkinson's disease. This allowed them to perform a virtual prevention trial, to determine whether anti-TNF therapy prevented the development of Parkinson's. They identified

144,000 individuals with inflammatory bowel disease and matched them with 720,000 unaffected individuals. They found that the incidence of Parkinson's among patients with inflammatory bowel disease was 20 percent higher than unaffected match controls. What was interesting is that there was a 78 percent reduction in the incidence of Parkinson's in patients with inflammatory bowel disease who had been given anti-TNF therapy. Thus, exposure to anti-inflammatory anti-TNF early in life was associated with a reduced incidence of Parkinson's disease. In a rat model of Parkinson's, administering a TNF inhibitor reduced loss of dopaminergic neurons. These findings support a role for systemic inflammation in both inflammatory bowel disease and Parkinson's disease. Inflammation is a known factor in inflammatory bowel disease, but discovering that treating inflammation could decrease the incidence of Parkinson's disease was a dramatic finding.

A PARKINSON'S GENE?

Genetics help provide another important insight into a major question in Parkinson's disease; namely, which patients develop cognitive problems with Parkinson's, and which do not? Treatment and prognosis can be drastically different if we know which Parkinson's patients will develop dementia and who will only have motor symptoms. Investigators found a clue in a gene called *GBA*, which is associated with Gaucher's disease, a rare genetic disease in which fatty substances build up in the liver and spleen. GBA is also a known risk factor for Parkinson's disease. It turns out that patients with Parkinson's disease who have GBA mutations have more cognitive decline. A study in Milan, Italy, classified 2,843 patients according to whether they had mutations in the *GBA* gene. The study found the risk of dementia was strongly affected by this type of mutation, which also affected survival. Brain imaging of these patients showed more widespread impairment than in those who did not carry a *GBA* mutation.

A study from Clemens Scherzer at our center examined 2,300 Parkinson's patients and also found a link between *GBA* mutations and the risk for cognitive decline.

These findings are in contrast to mutations of *LRRK2*, the other common genetic risk factor for Parkinson's disease. To date, it has not shown an effect on cognitive features of disease. It is not known how GBA mutations affect cognition, whether they relate to Lewy body pathology, synapse loss, or other issues. Because other factors may relate to cognition, testing for GBA mutations in clinical practice is under question. If somebody is GBA negative, it may provide false assurance to patients with the disease.

These issues are similar to what physicians and patients confront in preclinical Alzheimer's disease and whether they should be tested for the *ApoE4* gene, which is also linked to cognitive impairment. How much do patients want and need to know, especially if there are no direct therapeutic implications?

Identifying valid biomarkers and the ability to predict cognitive decline is a major challenge for the field. Because drugs that will be able to modify the disease are being tested, finding early signs of Parkinson's disease will allow treatments to begin earlier and enable us to also assess how well those therapies are working. Using the latest findings on GBA, Scherzer's group came up with a predictive model of cognition. They measured age of onset, baseline cognitive function, years of education, motor exam score, sex, depression, and *GBA* mutations to create a model that could predict cognitive decline in Parkinson's patients.

A major initiative to address biomarkers is called the Parkinson's Progression Markers Initiative, or PPMI, set up through the Michael J. Fox Foundation. Clemens Scherzer and Brad Hyman established the Harvard Biomarker Study to address this issue. Under the auspices of the PPMI, they followed 390 patients for two years and found five variables that could predict cognitive impairment two years later: older age, loss of sense of smell, sleep disturbances, imaging abnormalities, and spinal fluid markers.

Loss of smell and spinal fluid A-beta levels have also been reported as markers for Alzheimer's disease.

PRECISION MEDICINE

Two hundred years after James Parkinson identified the disease that now bears his name, we have entered the era of precision medicine. Precision medicine is defined as the customization of health care, with medical decisions, treatments, and products being tailored to the individual patient. Precision medicine has become the goal of therapy for virtually all diseases. This is nicely illustrated in breast cancer, where genetic mutations and tumor features determine the optimal therapy. Cohort studies of large numbers of patients are crucial for precision medicine. In our longitudinal CLIMB MS study of over two thousand patients, we have created a predictive model in which data from an individual patient is compared to all other patients in the database to help predict outcomes for that patient.

At the top of the list of factors that might be used in a precision medicine approach to Parkinson's disease is alpha-synuclein. If alpha-synuclein is at the core of Parkinson's disease, it is only logical that measuring alpha-synuclein could serve as a crucial biomarker, not only of disease but as a target for therapy. Because Parkinson's disease is not only confined to the brain, studies looking at skin biopsies of patients with Parkinson's showed they virtually all have alpha-synuclein in the skin. Higher levels of alpha-synuclein were found in skin biopsies of advanced Parkinson's and in those with disease of longer duration. Furthermore, a skin biopsy is a way to distinguish between Parkinson's and multiple system atrophy, which is also a Lewy body disease related to alpha-synuclein. Distinguishing the two conditions has been a major challenge. As it turns out, all Parkinson's patients who were tested had alpha-synuclein on the skin biopsy, whereas none of the patients with multiple system atrophy had it. Other tissues have been tested for alpha-synuclein, which has been found in the colon and in the salivary glands. In patients with Parkinson's disease, alpha-synuclein can also be found in the spinal fluid.

Interestingly, A-beta-42, the protein so important in Alzheimer's disease, predicts early-onset dementia and worsening of gait that is resistant to L-DOPA in early Parkinson's disease. In a group of Parkinson's patients who had a battery of spinal fluid markers measured, low baseline A-beta-42 was a negative sign implicating amyloid pathology that related to both dementia and L-DOPA-resistant gait progression in Parkinson's disease. It is possible that reducing the amount of amyloid in the brain could help some patients with Parkinson's disease.

The most accessible body component and fluid used for biomarkers is blood, and with sensitive techniques, it is now possible to measure blood alpha-synuclein. This could one day serve as a biomarker for Parkinson's. A more complicated blood test is to measure gene expression in the blood. This approach doesn't hypothesize what is abnormal—it simply looks for differences. Whole blood gene expression profiles were measured in Parkinson's patients and compared to those without Parkinson's, and researchers identified a gene signature that differentiated idiopathic (unknown cause of disease) Parkinson's disease from normal individuals: the expression of sixty-four genes was upregulated, and the expression of twenty-four genes was downregulated. The signatures revealed gene enrichment pathways that included metabolism, oxidation, and mitochondrial genes, pathways that have been postulated to be related to Parkinson's disease.

There is an inverse association between daily caffeine consumption and the risk of developing Parkinson's disease, predominantly in men. In women, there may be a protective effect, perhaps related to hormone therapy. Caffeine can protect against degeneration in dopamine-containing areas in the MPTP Parkinson's mouse model. Similarly, caffeine metabolites such as theophylline can attenuate MPTP neurotoxicity. This raised the question as to whether levels of caffeine and caffeine metabolites in the blood could serve as biomarkers of early Parkinson's disease. This was found to be the case but interestingly was not related to differing amounts of coffee consumption by the subjects tested. The levels of caffeine and nine metabolites of caffeine were decreased in patients with Parkinson's disease. Because these levels varied despite equivalent caffeine intake in Parkinson's disease patients and controls, it suggests there may be a difference in absorption or metabolism of caffeine in those with Parkinson's. These studies need to be replicated, but they provide an interesting window into potential serum biomarkers.

The most direct biomarker is imaging via PET scan or MRI, where the disease process in the brain is visible and can be used to follow disease progression. In MS, a classic MRI biomarker of the disease are spots that appear when a new disease area forms and light up when the disease is active. In Alzheimer's disease, PET imaging can directly measure amyloid and tau in the brain and is used to measure the effect of amyloid-lowering drugs. Unfortunately, because alpha-synuclein is primarily inside the cell in Parkinson's disease, there is no direct way to image alpha-synuclein in the brains of patients with the disease or measure the effect of treatment on alpha-synuclein in the same way that we measure it in the brains of Alzheimer's patients.

One can, however, assess dopamine function in the brain with PET compounds for the transmembrane dopamine transporter, or DaT. DaT scans are widely available clinically and commercially, and are used to measure the degree of dopamine loss and dopamine function in the substantia nigra. Sophisticated MRI imaging can also show changes in the brain, including anatomical loss and deposition of iron. The latest high-field MRI imaging is 7 Tesla, which can identify patients with Parkinson's from controls by determining the shape of the dopamine-rich substantia nigra. In MS, the 7 Tesla MRI allows us to get a better picture of disease on the surface of the brain and whether there is active inflammation.

Using imaging tests as a marker of disease progression, investigators measured free water in the substantia nigra. This provides a more precise assessment of substantia nigra structure. They found that changes in free water occurred in Parkinson's patients and not in controls; free water increased over the course of a year; and it was related to worsening over a four-year period. Thus, an increase in free water over one year could be used as a progression imaging marker for Parkinson's disease, which could be applied to clinical trials of disease-modifying therapies.

Other sophisticated imaging studies have been applied to measure disease progression and cognitive dysfunction. One example is the study of the physical shape and form of a structure called the nucleus basalis of Meynert, which has widespread connections to the rest of the brain and predicts cognitive decline in Parkinson's disease. Another example is imaging a structure called the locus coeruleus, which is associated with rapid eye movement sleep dysfunction in Parkinson's disease, which in turn is linked to cognitive deterioration. Advanced imaging of the locus coeruleus may reveal pathways that contribute to the high prevalence of non-motor symptoms in Parkinson's disease and will become important not only in treatment but in following the course of disease.

Although it is impossible to predict in a general population who will develop Parkinson's disease, those carrying genetic alterations such as the LRRK2 mutation are known to be at greater risk for Parkinson's disease. This is similar to the genetic mutations that are observed both in ALS and in Alzheimer's disease and creates the possibility for early or prophylactic therapy. In one study, prospective DaT imaging for dopamine was able to predict early development of Parkinson's disease in asymptomatic carriers of the LRRK2 mutation. Such studies will become crucial when specific disease-modifying therapies are available.

The most readily available biomarkers of Parkinson's are its clinical features, including cognition and motor and non-motor symptoms. It was generally felt that patients with a pure motor onset had a better prognosis. To address this question in a more detailed fashion, investigators attempted to classify clinical determinants of disease progression by subtype. Defining subtypes would be valuable in understanding Parkinson's underlying mechanism, predicting its course in an individual, and, of course, opening avenues to treating the patient.

The investigators used an approach called cluster analysis, which is now widely used throughout science. Specific features of a biologic system are put into a computer program that groups them into clusters based on their similarities. The authors used a prospective cohort of 113 patients with Parkinson's disease and put the data into the cluster program. They examined more than thirty variables that included a spectrum of features, including motor function, cognitive status, psychiatric manifestations, sleep disorders, and smell. Using cluster analysis, they were then able to identify subtypes.

One subtype was called "motor/slow progression." This cluster was characterized by slow progression of Parkinson's disease, in which tremor was slightly more prominent. Depression and anxiety were mild, and there was only minimal rapid eye movement sleep disorder, although mild cognitive impairment was present in almost half the group. The second subtype was called "diffuse/malignant." This cluster was characterized by mild cognitive impairment and sleep disorders. These patients had more severe motor symptoms and more depression and anxiety. The third was called "intermediate." This cluster had gait disturbances and were more likely to fall. Hallucinations were also common in this group.

The investigators then applied their criteria to the data from the Parkinson's Progression Markers Initiative and found that people with "diffuse/malignant" Parkinson's disease had the lowest level of CSF beta-amyloid and more brain shrinkage, while the milder "motor/slow progression" subtype had the least brain shrinkage. As expected, the "diffuse/malignant" type of Parkinson's progressed faster. These clinical and biomarker classifications can help clinicians in their care of patients and provide a basis for determining whether different subgroups of patients respond differently to disease-modifying therapy. This type of classification is done in Alzheimer's, where patients who are positive for the *ApoE* gene have more severe disease and may respond differently to therapy.

Like ALS and Alzheimer's, Parkinson's disease is classified as a progressive degenerative neurological disease. Unlike ALS and Alzheimer's, there is symptomatic treatment for patients with Parkinson's disease, and many patients are functional for many years after the diagnosis. Unfortunately, effective disease-modifying therapies are not yet available.

A RANGE OF TREATMENTS

Michael J. Fox's first symptom, trembling of the pinky of his left hand, began in 1990, when he was only twenty-nine years old. Over thirty years later he is still alive, and although he has many more symptoms of Parkinson's disease, Fox is married with four children and continues to function as an actor and advocate for the disease.

In 1999, Fox was waiting to testify at a Senate hearing on Parkinson's when he heard Gerald Fischbach, then director of the National Institute of Neurological Disorders and Stroke, say Parkinson's could be cured in five to ten years with the right funding and a little luck. That testimony inspired Fox to start a foundation for Parkinson's disease. It bears his name and has now raised over $750 million for Parkinson's disease. In those days, Fox was certain that a cure could be found in his lifetime. In a more recent interview with Jane Pauley, he stated that he didn't expect a cure in the next twenty years, but that within that time we would be able to improve quality of life for people with Parkinson's disease. With limitless funds, a cure for Parkinson's and other neurological diseases would occur faster. But how much money would it take to get us on the quickest road when we know that many discoveries happen by serendipity and that some cures will never be found until technology and basic science advance?

When Fox testified before Congress, alpha-synuclein was just being discovered. Clearly, the ultimate cure for Parkinson's disease will relate to preventing the toxic accumulation of alpha-synuclein. Although imperfect, the discovery of L-DOPA has been a major breakthrough in the treatment for Parkinson's disease.

In early stages of his disease, Michael J. Fox described taking a pill that was timed so that when he needed to act, he had the most positive "on" effect before the drug wore off. Treating Parkinson's symptoms involves careful management of the different medications to maintain appropriate therapeutic levels. Recent advances have involved gels administered to the intestine using a percutaneous pump. However, when the "on" and "off" effects became troublesome, Michael J. Fox chose an ablation procedure called a thalamotomy, which is an operation that destroys part of the thalamus in the brain to stop tremors. Thalamotomies are no longer performed for Parkinson's and have been replaced by deep brain stimulation, which is approved by the FDA and involves putting electrodes into the brain by stereotatic surgery.

The current treatment for Parkinson's disease is targeted at ameliorating symptoms with no effective means to modify the disease course. There are complicated algorithms for treating Parkinson's disease, which center on drugs affecting the dopamine system in the substantia nigra. The drugs include levodopa, carbidopa, peripheral decarboxylase inhibitors, monoamine oxidase inhibitors, and dopamine agonists. L-DOPA has the greatest anti-Parkinsonian effect for motor signs and symptoms, including tremor, rigidity, and slowness of movement, with the fewest adverse effects in the short term; however, long-term use is associated with motor fluctuations and dyskinesias, or abnormal movements. Physicians debate when L-DOPA therapy, as opposed to other ways of affecting the dopamine system—through dopamine agonists or monoamine oxidase inhibitors, for example—should be used.

Non-motor symptoms can be difficult to treat. These include nausea, sleep disturbances, depression, hallucinations, low blood pressure that causes dizziness when standing up, impulsive behaviors such as gambling and hypersexuality, and dementia. Most devastating, of course, is dementia.

Natural or herbal medicine is often touted as a treatment option. Mucuna pruriens is a levodopa-containing leguminous plant that grows in all tropical areas. Its seeds contain high concentrations of L-DOPA. In rural Africa, only 15 percent of patients with Parkinson's disease are treated with L-DOPA, and some find benefit from mucuna pruriens seeds. Studies have been performed comparing mucuna pruriens seeds with L-DOPA and L-DOPA plus carbidopa combinations, and the seeds were found to be as effective, with acceptable side effects. When it comes to herbal or natural medicines that have been indicated as beneficial to disease, it's important to know that these treatments could contain unknown substances or drugs. That being said, a Chinese colleague of mine who worked on sophisticated T cells in MS continues pharmaceutical development in China, part of which is the search for natural remedies with traditional Chinese medicine.

Psychosis is one of the non-motor symptoms of Parkinson's disease that illustrates how widespread Parkinson's is in the brain. Early in the disease, patients may experience what are called "passage hallucinations," when a person, animal, or indefinite object is seen briefly passing in the peripheral visual field. Other abnormalities are illusions in which faces or objects are seen in different forms, such as appearing as clouds. Later, formed visual hallucinations of animals or people occur in which the patient realizes they are hallucinating. As the disease progresses, the patient experiences hallucinations without knowing that they are hallucinations, and, finally, has delusions. It appears that the progression of these abnormalities are connected in many ways to the progression of the disease according to the Braak stages (see page 230), with the most severe psychosis associated with widespread Lewy body pathology in the brain cortex. Certain genetic links, including GBA, have been identified as being associated with Parkinson's and psychosis.

The treatment of Parkinson's disease symptoms usually provides good control of motor abnormalities for four to six years, after which on/off phenomena and abnormal movements may appear. Levodopa/carbidopa is the gold standard of symptomatic treatment. Monoamine oxidase inhibitors may be given early in the disease. Other drugs that stimulate dopamine, such as ropinirole and pramipexole, can be given as single therapy early or later in moderate to advanced disease. Drugs that affect the cholinergic system, such as Cogentin, are second-line drugs used for tremor only. Other drugs are given for constipation, daytime sleepiness, blood pressure changes, and depression. Some symptoms can be related to dopamine replacement therapy, including psychosis, visual hallucinations, and impulse control, though they also occur unrelated to treatment when there is widespread damage to the brain. Some neuropsychiatric symptoms occur in Parkinson's disease independent of the loss of dopamine in the striatum.

Exercise and physical therapy may also play an important role in Parkinson's disease. Formal studies have evaluated different types of exercise, and it appears that a minimum of four weeks of gait training or eight weeks of balance training can have positive effects that can last as long as a year. Strength training, aerobic training, tai chi, and dance therapy are also effective, and may even increase the response to medications. We've established a similar program for MS patients. As a physician, one always wants to provide positive routes to health, and exercise is clearly one of them, although it won't provide a cure. New technology includes laser shoes in which lasers attached to each shoe give patients visual cues about where to go next. Other technology includes Google glass and wearable devices that can help monitor motor function and response to therapy. But not everything that would appear to help does. We know that caffeine may have a neuroprotective effect, but a small trial of caffeine did not show a positive effect on motor manifestations of Parkinson's disease.

Falls are frequent and are a serious complication of Parkinson's disease, related, in part, to a defect in the systems that contribute both to gait and cognition. Rivastigmine is an anticholinesterase inhibitor that is licensed for the treatment of Parkinson's disease dementia. A Phase 2, double-blind trial suggested that rivastigmine can improve gait stability and might reduce the frequency of falls in patients with Parkinson's disease. We often look to currently prescribed drugs to see if they can provide benefits that were not initially anticipated.

Although the mainstay of treatment for Parkinson's disease is based on L-DOPA, treatment does not rescue the dopamine pathway. Furthermore, dopamine-based treatments lose efficacy over time and may be limited by side effects such as hallucinations and abnormal movements. Because of these limitations, investigators began examining the possibility of using cells as a dopamine replacement strategy. The very first experiments in which cells were transplanted into the brain took place over a century ago, but the experiments that ushered in the modern area of grafting cells in the brain began in the 1980s in Sweden.

In 1982, two Swedish patients with Parkinson's disease had tissue taken from their adrenal gland and grafted into their brains near the caudate nucleus. Although there were no major clinical benefits, two additional Swedish patients were transplanted with tissue from their own adrenal glands in 1985, this time into a different part of the brain called the putamen. These patients experienced short-lived improvements in motor function. Olle Lindvall, part of the team that did the second graft, recalled presenting the findings to the New York Academy of Sciences in 1986 and receiving an "overly negative" reaction.

However, professional opinion of the brain grafts changed dramatically in 1987 with a publication in the *New England Journal of Medicine* that showed transplanting tissue into the brain had major benefits in two Parkinson's patients in Mexico. One of the patients had been "unable to perform even the most basic activities." Five months after surgery, he could walk and eat without help. He could even play soccer with his five-year-old son. The second patient was also severely disabled, beset with uncontrollable tremors. Within three months after surgery, the tremors vanished almost completely.

Following a positive editorial in the *New England Journal*, several hundred patients underwent the procedure. Of the cases with careful scientific follow-up, none experienced the dramatic results of the original two patients. It also became clear there were complications from surgery, including psychiatric disturbances, and when people died, there was no evidence of good graft survival. Overall, the long-term outcome of the surgery was poor. This was before the era of deep brain stimulation, which helps with motor system dysfunction and, as we describe below, has become an FDA-approved treatment for Parkinson's disease.

Despite the questionable success of the early grafts, these initial treatments paved the way to grafting dopamine brain cells from the ventral mesencephalic brain region of human fetuses. They were called FVM transplants. These cells had positive effects in animals, and in 1987 the first FVM transplants in Parkinson's patients were performed in Lund, Sweden. Although the first two patients did not improve, the next two did, both clinically and in brain imaging. Following this success, another thirteen patients had grafts in the 1990s, again in an open-label fashion—they all received human FVM tissue, prepared from fetuses with a gestational age of six to eight weeks. Immunosuppression was given to prevent rejection. Although not everyone who received a graft was helped, there were some patients who no longer required medicine, and imaging showed restoration of normal dopamine signaling. This led to many other studies, with variable results.

In 1993, when President Clinton allowed funds to be made available for FVM transplants, two NIH-funded studies were carried out. It would be the first double-blind testing of transplantation, which some felt could not be done. Some patients underwent surgery and others sham surgery. Unfortunately, to the surprise of the investigators, these two double-blind trials showed no benefit when all groups were compared. In addition, half of the patients developed abnormal movements, perhaps related to stimulation of the serotonergic neurons in the brain. It also was concluded that human FVM transplants did not significantly improve the clinical status of patients with Parkinson's disease, especially with newer therapies available, such as deep brain stimulation. Grafting tissues derived from other sources was attempted, but none showed a positive effect. Meanwhile, in Europe, the European Union has funded TRANSEURO, which is an ongoing open study of twelve patients who will receive human neurotransplants.

A parallel effort is GForce-PD, in which groups will study human embryonic stem cell lines to see if dopaminergic neurons can be generated from them. Even more exciting is the ability to induce pluripotent stem cells that can make dopamine. This technology means that it is possible to take a skin biopsy or blood cells from the patient and transform them into dopaminergic neurons. Investigators have recently shown that human dopamine pluripotent stem cells functioned well in a primate model of Parkinson's disease. They investigated dopaminergic neurons derived both from patients with sporadic Parkinson's disease and from healthy individuals, following the animals for over two years, with positive results.

Induced pluripotent stem cells may be the wave of the future if cell transplants are to work. An even higher-tech approach would be to reprogram the cells in the patient, rather than perform cell replacement surgery. Using genetic techniques and understanding transcription factors, investigators were able to reprogram human astrocytes in the test tube and show that they could produce dopamine. Even more exciting, in a mouse model of Parkinson's disease, these transcription factors, when injected into mice, corrected some abnormal motor behavior in the animals. This approach raises the theoretical possibility that clinical therapy for Parkinson's disease could ultimately be undertaken by delivery of genes rather than cells.

The story of cell transplantation in Parkinson's disease is a cautionary tale. Treatments that appear to work spectacularly well may be shown to not work at all when they are studied with scientific rigor. Hopefully, as technology advances, we will enter an era with sophisticated patient-derived stem cells. Unlike the stem cell transplants now offered to Parkinson's patients, who must become medical tourists in search of a cure, these will be based on rigorous animal testing, with formal follow-up, and they could be a great benefit to Parkinson's patients.

Unlike the heretofore disappointing failure of stem cell brain implants, deep brain stimulation—stimulating the brain with electricity, a conceptually low-tech approach—has been found to help those with Parkinson's disease. At age sixty, John Rogers developed a tremor in his right hand. He first noticed it playing golf. His symptoms progressed over a year, at which time he was diagnosed with Parkinson's disease.

Over the next decade, he was treated with many drugs. Initially they helped, but they became less effective over time. He took all available drugs that targeted dopamine but had troublesome involuntary movements on his right side, although he was taking one-and-a-half tablets of carbidopa/

levodopa every two hours. The medication helped with his rigidity and walking, but he experienced six hours of "off" time, when his symptoms were unresponsive to the medication. In addition, he had severe disabling involuntary movements for four hours a day. He had mild symptoms on his left side that did not really bother him. He was otherwise healthy and had no problems with cognition, continuing to work as a lawyer. He decided to undergo deep brain stimulation.

Rogers was hospitalized and taken to the operating room, where a hole was drilled in his skull and electrodes were implanted deep in his brain in areas that are important for coordination of movement. Before the procedure, he was taken off his medicine and was awake as the electrodes were positioned. Once the electrodes were in place, the electricity was turned on and he underwent deep brain stimulation. His involuntary movements went away immediately. A wire was attached to the implanted electrode, tunneled under the skin of his scalp to his chest, and connected to an impulse generator.

Both the brain and the heart depend on electricity. There are many types of cardiac pacemakers, but there are few instances where stimulating the brain makes a difference in disease. It works in Parkinson's. Deep brain stimulation inhibits the firing of some of the nerve cells and the excitation of the neighboring parts of the brain. In addition, the astrocyte cells in the brain are stimulated to release calcium, which helps cell metabolism and possibly increases local blood flow.

Symptoms responsive to L-DOPA, such as tremor, on-off fluctuations, and abnormal movements, are most likely to improve with deep brain stimulation, whereas other symptoms, including gait, balance, and speech, are less likely to improve. Most prospective, randomized studies, however, report a slight cognitive decline following deep brain stimulation. Four randomized, controlled trials of deep brain stimulation found positive results, and the procedure is FDA approved. That gives patients like John Rogers another option when their medications lose effectiveness. Nonetheless, deep brain stimulation, like L-DOPA, only targets the pathway related to motor function, not the multiple areas containing abnormal deposits of alpha-synuclein that are the underlying cause of the disease. Although symptomatic treatment such as L-DOPA and deep brain stimulation can help people function better, the ultimate goal is to find a disease-modifying therapy—treatment that interferes with the underlying processes that drive the disease.

There are a number of drug targets in Parkinson's disease that are designed to modify the disease process rather than provide symptomatic therapy that targets dopamine pathways. One such target is calcium. Calcium plays a central role in the normal functioning of neurons, and calcium channels are prominently expressed in brain areas affected by Parkinson's disease. The selective vulnerability and degeneration of dopamine neurons in Parkinson's disease could be related to the high-energy demands caused by influx of calcium into a cell, which then leads to oxidative stress and cell death. Indeed, there is epidemiologic data that patients treated with calcium channel blockers may have a reduced risk for developing Parkinson's disease.

A drug called isradipine has a high affinity for calcium and is an FDA-approved drug for treating hypertension, illustrating the potential to treat Parkinson's disease with drugs that are FDA approved for other diseases. In animals, isradipine treatment affected dopamine neurons in the brain and protected them from oxidative stress. These encouraging animal studies led to a randomized, controlled trial of isradipine in patients with Parkinson's. The study was too small to show a clinical benefit, but dosing and toxicity were established and supported trying it in a large trial.

A Phase 3 trial called STEADY-PD III was initiated and enrolled 336 people at fifty-five sites in less than a year. It was an expensive study, funded by a $23 million NIH grant. The people who entered the three-year study had not been receiving standard therapy with dopamine drugs and were given either the study drug or placebo. Study measures included time to initiation of L-DOPA therapy and time to severity of motor complications, with the primary outcome being change in the Parkinson's disease rating scale. Unfortunately, the three-year trial was negative, a disappointment for all involved.

As discussed, an important factor driving neuronal death in Parkinson's disease is related to oxidative stress, an imbalance between toxic free radicals and antioxidant defense mechanisms. In fact, the crime scene shows evidence of oxidative damage in the substantia nigra of patients with Parkinson's disease. Oxidative damage causes failure in the neurons' normal protective mechanisms, and it changes dopamine metabolism. Several drug trials have been carried out to reduce oxidative stress in Parkinson's disease. These include the antioxidants selegiline, rasagiline, vitamin E, coenzyme Q10, mitoquinone, and creatine. Unfortunately, none have shown positive results. It may be that although there is oxidative damage, these antioxidants are too little, too late, or not strong enough to have an effect.

Nonetheless, according to epidemiologic data, people who have high levels of serum urate, which is a natural antioxidant, have a third less chance of developing Parkinson's disease than people with normal levels of urate. It also appears that people who have high levels of urate in the serum or the spinal fluid who have early Parkinson's disease may have a reduced rate of clinical progression. Inosine, a precursor of urate, was found to prevent degeneration of dopamine neurons in an animal model of Parkinson's disease. This led to randomized clinical trials of oral inosine. Although initial trial results were not particularly positive, a larger, Phase 3 trial is being considered. However, an important caveat is that increasing serum urate could also increase the risk of high blood pressure, heart disease, gout, and stroke over the long term. This poses a potential problem for older patients and illustrates that there are complications and off-target side effects of medications, making it hard to find drugs to treat the elderly population.

Iron accumulation in the brain occurs in Parkinson's disease and can be seen on MRI in the basal ganglia. Although not a primary cause of Parkinson's disease, iron accumulation could contribute to damage. Interestingly, iron accumulation also occurs in later stages of MS. Animal studies support the neuroprotective effect of lowering iron using chemicals called chelating agents, which are used to remove toxic metals from the body. The chelating agent deferiprone showed some positive effects in a recent trial. However, the effects were not long lasting; further Phase 2 and 3 trials are now underway.

Considerable evidence from many areas of the crime scene, including genetics, animals, and autopsy suggest mitochondria may be dysfunctional in Parkinson's disease. Mitochondria are the energy machines of the cell. The selective vulnerability of dopamine neurons could be related to their high energy and metabolic demands. However, recent trials using mitochondrial enhancers such as creatine and mitoquinone have not been successful. High-throughput screening using cultured nerve cells from patients with Parkinson's disease revealed that ursodeoxycholic acid can rescue mitochondrial dysfunction and is a candidate for clinical trials.

Glutathione is involved in neurotransmission between nerve cells and is also a primary antioxidant of the brain. Glutathione depletion is seen in the substantia nigra of patients with Parkinson's disease and is part of the crime scene. Intravenous administration of glutathione was not successful, perhaps because it couldn't get into neurons. However, a compound called N-acetylcysteine, an oral precursor of glutathione, decreases alpha-synuclein aggregation in mouse models of Parkinson's disease. It hasn't been tried yet in Parkinson's patients, but it didn't work in early Alzheimer's disease. To be effective, it would best be given very early, ideally in presymptomatic stages of the disease. This raises the conundrum that we should be treating Parkinson's disease as early as possible but can't definitively identify early or presymptomatic patients.

As discussed, drugs used for treating other diseases may have application to Parkinson's disease. One such drug is pioglitazone, which is FDA approved for type 2 diabetes. It provides a degree of neuroprotection in animal models of Alzheimer's disease, stroke, and ALS, and is currently being tested in Parkinson's disease. Drugs such as statins not only lower cholesterol levels but also have marked anti-inflammatory and neuroprotective effects. Simvastatin can reduce alpha-synuclein aggregation. Some studies suggest that taking statins may provide protection against Parkinson's.

However, a recent population study looking at insurance claims for more than fifty million people identified those with Parkinson's and found that the use of statins was associated with a *higher* risk of developing Parkinson's disease, especially in those using statins for less than two and a half years. Conflicting data may have been related to an independent effect of cholesterol on Parkinson's disease. Of note, simvastatin has been reported to help progressive MS.

Several drugs approved for other neurodegenerative diseases have been tried in Parkinson's disease. Riluzole, a drug approved for the treatment of ALS, was tested in Parkinson's disease without effect. Minocycline, an antibiotic with anti-inflammatory properties that helps in early MS, also showed no effect in Parkinson's. Interestingly, in ALS the antibiotic minocycline may have made the disease worse. Minocycline presumably affects microglia, and microglia can have both positive and negative influences in the brain. Although there are common mechanisms among all the diseases that affect the brain, there are also variables. In animals, we have found some treatments that make ALS better but make Alzheimer's disease worse.

Glucagon-like peptide 1 receptor (GLP-1R) drugs are used to treat diabetes and act on pancreatic beta cells to decrease glucose levels. They may also work to combat neurodegeneration in Parkinson's disease. The GLP-1R agonist drug exenatide has shown positive results in a number of models of neurodegeneration, and a clinical trial in Parkinson's patients given daily injections of exenatide for twelve months showed improvement in both motor and cognitive function. More studies are needed, and there are other drugs in this class that could be tried. Exenatide illustrates an important path forward in finding drugs for Parkinson's disease.

One of the easiest treatments readily available is nicotine, as there have been numerous studies suggesting smoking helps patients with Parkinson's disease. There is even some scientific evidence that in certain models it might affect dopamine neurons or oxidative stress. However, human trials in which high-dose transdermal nicotine was given to Parkinson's patients have not shown benefit.

TRIAL AND ERROR AND POTENTIAL SUCCESS

Over the years, a great deal of interest has focused on trophic factors that can affect dopamine cells. Trophic factors are small proteins that control the growth, differentiation, and survival of neurons. Studies have shown a number of trophic factors support the viability of dopamine neurons, with glial-derived neurotrophic factor, or GDNF, getting the most attention. Gene delivery of GDNF in animals prevented degeneration of these dopaminergic neurons. The success in animals prompted Amgen to perform a clinical trial, testing the safety and efficacy of GDNF, which was directly infused into the brains of patients with Parkinson's disease. Although there were no side effects, the trial reported no efficacy, and when the brain of a patient who died was examined, there was not good penetration of the trophic factor into the brain. Enthusiasm for GDNF was renewed when gene therapy enabled good delivery of GDNF into the dopamine area of the brains of monkeys. Because of this, both open and double-blind trials assessing gene therapy with GDNF were carried out. Although there were positive clinical signals in the open trial, the subsequent double-blind trial was not positive, reinforcing the fact that open studies that do not control for placebo effect and bias are often misleading. The company Ceregene carried out two Phase 2 gene therapy trials with AAV-neurturin, a sister molecule of GDNF. Unfortunately, both Phase 2 trials also failed. It may well have been that the disease was too far advanced for these factors to work.

A new target for the treatment of Parkinson's disease is NURR1, a hormone receptor known to play an important role in dopamine neuron development differentiation and survival. Importantly, it has been shown that defects in NURR1 are associated with Parkinson's disease. Ablation of NURR1 in mice leads to behavioral features similar to Parkinson's disease. Other studies show that NURR1 gene therapy cannot only enhance dopamine transmission but also protect dopamine neurons from cell injury induced by environmental toxins or inflammation from microglia.

Investigators have devised compounds that interact with NURR1 and have the dual effect of neuroprotection and reducing symptomatic effects in disease models by preventing dopamine neuron demise.

A major current focus for the therapy of Parkinson's disease is targeting the *LRRK2* gene. Although Parkinson's disease has not been considered a genetic disease, approximately 2 percent of Parkinson's populations have a mutation in the *LRRK2* gene, with some populations having an even higher percentage. As discussed previously, LRRK2 is thought to work as a molecular switch, turning other proteins in the cell on and off. As we know from alpha-synuclein, proteins do not function normally in the neurons of patients with Parkinson's. More importantly, the clinical picture of a person who has the *LRRK2* gene mutation resembles Parkinson's disease in a young person and sporadic Parkinson's disease in an older person.

It is believed that LRRK2 is a target for a wide range of people with the disease and may play an important role in sporadic disease. The mutations in the *LRRK2* gene cause the gene to be overactive and toxic to the cell. Although the *LRRK2* gene was found to be linked to certain forms of Parkinson's disease in 2004, it took almost a decade-and-a-half for first-in-human dosing of a compound that targets LRRK2. Denali Therapeutics accomplished this in 2017, testing the safety of their LRRK2 inhibitor in healthy volunteers. The Michael J. Fox Foundation for Parkinson's Research helped foster collaboration among different drug companies and academic scientists to move this forward. Indeed, many drug companies now have LRRK2 inhibitors in their pipeline.

The story of finding a gene, understanding its function, determining how it exists in populations, finding a way to target it, and then moving it into people is slow, laborious, and expensive. When patients come to the doctor, they always want to know how soon there will be a cure. The story of LRRK2 illustrates how long that path can be. If successful, it will take another five to ten years and as much as a billion dollars to develop safe LRRK2 inhibitors.

Of all the targets for treating Parkinson's disease, alpha-synuclein is probably the most exciting. Alpha-synuclein is directly related to the disease because Lewy bodies contain alpha-synuclein deposits. As previously mentioned, in every disease, there is a sequence of events, which are characterized as "upstream" and "downstream" in the disease process. For example, in poliovirus, the most upstream event in the disease is infection. Once the virus causes damage, downstream events are harder to treat. Most of the processes that occur in Parkinson's, including oxidative stress and mitochondrial dysfunction, are downstream of the toxic effects of alpha-synuclein. The conversion of alpha-synuclein from its natural form to the toxic aggregated form has been established as the key initiating event in the cause of Parkinson's disease. Clearly, if one can stop the toxic accumulation of alpha-synuclein, one could stop the disease.

One of the major approaches used currently to clear alpha-synuclein is treatment with antibodies against alpha-synuclein, similar to what is being attempted in Alzheimer's disease. In addition, other approaches to clear alpha-synuclein from the cell are being developed that involve targeting specific pathways inside the cell that interfere with alpha-synuclein and thus decrease its toxic effects in the cell. Dennis Selkoe is searching for compounds that will prevent the breaking up of the normal four-stranded complex of alpha-synuclein that was discussed previously. Of course, if the goal is to decrease alpha-synuclein, a major breakthrough would be to develop a test to show the treatment is working. With Alzheimer's, PET imaging can show A-beta in the brain, and imaging is used to track antibodies that are designed to lower A-beta levels. Unfortunately, to date, there are no methods for imaging alpha-synuclein in the brains of Parkinson's patients.

CURING PARKINSON'S DISEASE

In 2017, the scientific community marked the two hundredth anniversary of James Parkinson's description of the disease that took his name. In those two hundred years, a great deal of progress has been made, but we are still a long way from a cure. Major advances involved understanding the role of the dopamine system, and we have a number of drugs that help Parkinson's patients by acting on dopamine. However, these drugs do not affect the course of the disease. We now know that Parkinson's is a degenerative disease that impacts more than the dopamine system. Parkinson's is related to an interplay of both genetic and environmental factors. At least fifteen causal genes and more than forty susceptibility gene loci have been identified. Identification of alpha-synuclein in Lewy bodies has led to trials of immune approaches to clear alpha-synuclein. We know that many specific problems in the cell, including toxic oxygen species and mitochondrial dysfunction, all contribute to Parkinson's disease. Furthermore, other cells such as microglia play a role by causing inflammation and promoting the disease, as they do in Alzheimer's disease.

It is unlikely that we will see a headline announcing a single breakthrough in Parkinson's disease. More likely, as in cancer, Parkinson's will ultimately be cured by carefully testing and adding new drugs and different combinations of drugs to the arsenal. The connections between Parkinson's and other diseases, such as Alzheimer's, and modulation not only of alpha-synuclein but also of A-beta may be important. Phase 3 clinical trials of promising compounds are underway that will provide more data about the disease. At the moment, much of what has been done has moved us toward neuroprotection—protecting neurons from death or dysfunction that occurs not only in Parkinson's disease but also in ALS, Alzheimer's disease, and progressive forms of MS. Of course, people with these neurological diseases would much prefer neuro-rescue, in which a deceased neuron could somehow be restored, which is a far more difficult task.

In the two hundred years since Parkinson's was identified, we have made a lot of progress in better understanding the disease. Will it take an additional two hundred years to find a cure? I do not believe it will take that long, although it will be more than five or ten years. I think we are closer to the goal, although Gerald Fischbach of the NIH said the disease would be cured in five to ten years when Michael J. Fox testified before Congress almost two decades ago.

We will get there because we now have tools to better understand the disease, such as improved animal models and the ability to study human-induced pluripotent stem cells, which will allow investigators to generate new targets for treatment. In the past, clinical trials did not effectively stratify Parkinson's patients, and we now have better ways to treat different subsets of patients. We will certainly need to begin treatment at earlier stages of the disease if we are going to cure it. The emerging concept of prodromal Parkinson's disease will be key, and the development of biomarkers that can identify people at risk or who have early disease will be crucial.

Of course, the central question is, How does Parkinson's really begin? What are the origins? Non-motor symptoms such as sleep, smell, and constipation need to be better understood and linked to blood or urine biomarkers or imaging. Another central question relates to alpha-synuclein: Will targeting alpha-synuclein and targeting it early affect the disease? It is hoped that alpha-synuclein targeting therapies will be one of the major breakthroughs in Parkinson's disease, just as there are hints that targeting A-beta may help in Alzheimer's disease. What is missing in Parkinson's disease are imaging techniques that will enable us to measure the removal of alpha-synuclein. Another big question relates to the role of the gut in Parkinson's disease. Initial studies of the microbiome and the fact that constipation is often an early sign of Parkinson's suggest that this brain disease may indeed start in the gut. If there is a breakthrough, it could theoretically be identifying the factors in the gut that trigger the disease, factors that may be treatable or preventable.

The cure for Parkinson's disease will come when we have truly identified the upstream target—which we now believe to be alpha-synuclein—treat that target, and measure the effect of treatment with validated markers to monitor disease progression. In addition to biologic measures, our measurements may include wearable devices and big data. We will need to separate subtypes of patients, screen those at risk, and understand prodromal Parkinson's disease.

Finally, the days of scientists looking into a microscope by themselves are over. Major funds are needed to fund collaborative research. In 2018, the Accelerating Medicines Partnership-Parkinson's Disease was launched. Known by the acronym AMP-PD, the partnership forges cooperation among the government, the pharmaceutical industry, and private organizations such as the Michael J. Fox Foundation. These groups can work together harmonizing results, generating genetic data, and building large-scale biomarker discovery programs. There is also a new global basic research initiative to uncover the roots of Parkinson's, called ASAP (Aligning Science Across Parkinson's). One must also keep in mind that there may be a scientist who has never heard of Parkinson's disease, working on a gene or protein in a basic laboratory, perhaps studying flies or fish, who could discover the key that has a major impact and leads to curing the disease. We can never forget or underestimate the importance of basic research.

CHAPTER SIX
Glioblastoma

In 1998, I was sitting in a conference room at the old Peter Bent Brigham Hospital, discussing our plans to establish an MS center that would include an infusion room, MRI, and special clinical facility. Robert Bretholz, a hospital trustee who supported our research, was at the meeting and excused himself to go to the washroom. Concerned, I followed him. I knew Rob had been diagnosed with brain cancer. When I caught up with Rob, he said, "I had to step out for a few minutes, Howard. These tiny seizures are killing me."

I knew Rob well because I had helped care for his wife, who had MS. I had met them fifteen years earlier when this vibrant, young, handsome couple appeared in my office. I remember telling my wife how they reminded me of movie stars. This was the first chance I'd had to be alone with Rob to talk about his brain cancer.

"I came through the surgery okay," he said, "but the radiation is making me tired."

"How did it start?" I asked.

"A seizure," he said. "I had been having some headaches, but I didn't pay attention to them. But after the seizure, they saw it on the CAT scan and MRI."

"How do you feel?" I asked.

"I'm feeling okay and haven't missed a step with my business and hospital activities. I'm going to fight this, Howard," he said. I looked at him and said nothing. "What do you think my chances are?" he asked.

I paused. "It's an uphill battle," I said finally. "But I think you have a good chance of making it." I didn't have the heart to tell him that his outlook was dismal.

One year later, I attended his funeral.

Glioblastoma is one of the deadliest forms of cancer and one that has features unique to the brain. It is the most common primary malignant brain tumor in adults. *Primary* means the tumor begins in the brain, as opposed to spreading from another part of the body. When I saw Rob, the median survival time was about a year from the time of diagnosis. It's now fifteen months.

Many well-known people have had a glioblastoma. Senator Ted Kennedy survived for fifteen months, composer George Gershwin for twelve months, singer Ethel Merman for ten months, political consultant Lee Atwater for twelve months, and Beau Biden, Delaware attorney general and son of President Biden, twenty-one months. Recently, we also lost John McCain to a glioblastoma after twelve months.

Cancer has been known since the dawn of history. Fossilized bones in Egyptian mummies show evidence of tumors. Ancient Egyptian papyrus manuscripts describe breast cancer and noted "there is no treatment." The word *cancer*, coined by Hippocrates, the Greek physician, comes from the Greek *karkinos*, a giant crab in Greek mythology. He thought an imbalance of fluid and an excess of black bile in a particular organ caused cancer. This was the prevailing theory through the Middle Ages; subsequent theories blamed fluid called lymph, chronic irritation, trauma, and parasites.

Once DNA was discovered by Watson and Crick in 1953, it became clear that cancer is caused in large part by damaged DNA. In addition, the high incidence of breast cancer in nuns compared to women who have been sexually active provides a hint that hormone receptors on breast cancer cells may play a role in cancer. A better understanding of cancer has brought new and better treatments. Surgery, radiation therapy, and chemotherapy have all added to our success in the treatment of cancer, and the ability to make monoclonal antibodies has led to tools to treat cancer by modulating the immune system. Therapies now exist to successfully treat many cancers, transforming them from killers to a disease that can be cured. Unfortunately, glioblastoma is not one of them.

THE CRIME SCENE

Let us begin by taking a careful look at the crime scene to see what makes glioblastoma so deadly. A doctor's first view of the crime scene comes when they view an MRI scan that shows an irregular mass in the brain. The mass has uneven edges, giving it an ugly, moth-eaten appearance. There is often swelling around the tumor. The rim of the tumor often lights up when gadolinium dye is injected into the bloodstream. The center of the tumor is dark, almost like a black hole. Some investigators have identified different subtypes of glioblastomas on MRI, which may be linked to the molecular pathways of the tumor, and whose clustering can help predict survival. Such MRI clusters are classified as multifocal, spherical, and rim-enhancing.

Viewing the crime scene under the microscope may show tumor cells throughout the brain. These deeply infiltrating tumor cells are more likely to escape surgery, and it may be that the infiltration is the property of a more resilient and difficult-to-treat population of cancer cells that initiates and drives tumor recurrence. Because the tumor is characterized by infiltrating growth, during surgery the tumor mass is not always distinguishable from normal tissue. When looking at a tumor under the microscope once it has

GLIOBLASTOMA

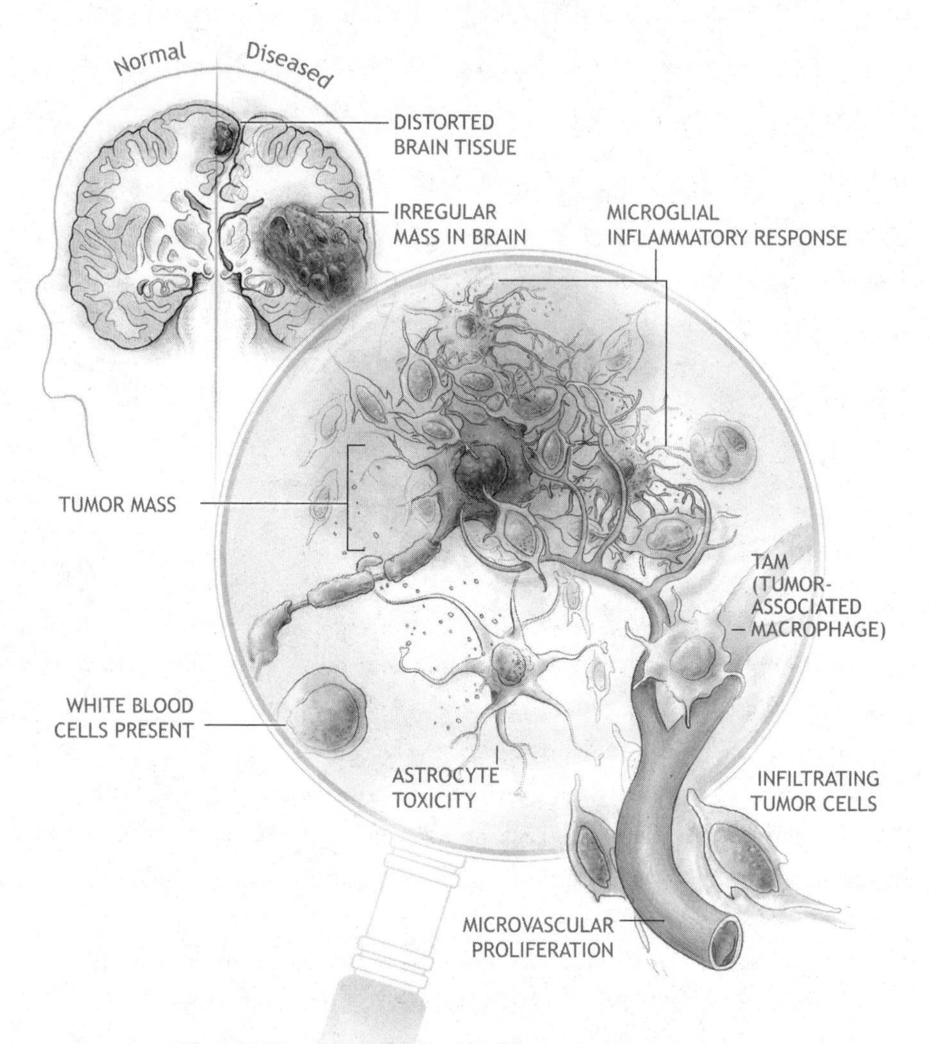

The glioblastoma crime scene shows an irregular tumor mass in the brain. Tumor is characterized by infiltrating growth with varied disorganized cells. White blood cells infiltrate, and damaged blood vessels form clusters. Microglial cells and astrocytes are also affected. In summary, it is an ugly, wild picture.

been removed, what is most striking is how varied the appearance of the tumor is and how disorganized the cells are.

After looking at the crime scene from the outside with MRI, via surgery, and under the microscope, let's go deeper into the physiology and examine the cellular and molecular fingerprints at the crime scene of glioblastoma. The tumor consists of small cells of many shapes, with large nuclei and coarsely clumped particles in the cell. Gene profiling has stratified glioblastoma into four major molecular subgroups: neural, proneural, classical, and mesenchymal. There are different subtypes of tumor in different landscapes of the crime scene, which may respond differently to therapy. The story gets more complex, as these subtypes can all exist in the cellular subpopulations of a person's individual tumor. White blood cells are also found there, including lymphocytes, neutrophils, and macrophages. As much as one-third of the cells in the tumor are macrophages, the scavenger cell of the immune system.

A classic feature of the crime scene is the presence of blood vessels, which form clusters and look a bit like the kidney structures that filter blood into urine. There are blood clots that cause damage to the vessels; large areas of cell death within the center of the tumor or irregularly shaped areas that contain dead cells; and cells that are arranged like wooden fence posts on the American frontier, which represent rapidly growing cancer cells that have outgrown their blood supply. In summary, the scene of the crime is an ugly, wild picture.

Virtually all glioblastoma patients are treated with surgical removal of the tumor, called a resection, so that we have a detailed view of the crime scene in a living patient, something we don't have in other neurological diseases. The tumor itself is an amorphous mass that pushes aside normal brain tissue. The blood vessels make it look like luxuriant vegetation, rich nutrition for the tumor, but there are also areas of complete death of tissue, where it looks like a desert.

Another feature of the crime scene relates to metabolic changes. Neural cells are critically dependent on glucose for survival, and tumor cells rapidly consume glucose. Paradoxically, malignant brain tumors are less common in people with high blood sugar or diabetes. A three-year study led by David Nathanson at UCLA found that disrupting the tumor's glucose intake in mice prevented the glioblastoma from getting bigger. Addressing metabolism may be important to ultimately improving our current treatment strategies for glioblastoma.

Picturing a tumor or cancer, one might imagine tumor cells growing wild in the brain, like weeds in a garden. But as it turns out, as many as 30 to 50 percent of the cells in the glioblastoma are not tumor cells at all but white blood cells called tumor-associated macrophages (TAMs). These macrophages enter from the bloodstream and create a fertile garden for the growth and expansion of tumor cells—after being recruited to the tumor, they release factors that help tumor growth and survival. They can also serve to prevent the immune system from coming in to destroy the tumor. Also inhabiting the crime scene are brain microglia, those Dr. Jekyll and Mr. Hyde cells we discussed previously.

ORIGINS

Amazingly, there are few genes or other factors that are strongly linked to glioblastoma. Unlike ALS or Alzheimer's disease, where there are dominant genes that cause the disease, there are no dominant genes that cause glioblastoma, though there are rare genetic disorders linked to glioblastoma, such as neurofibromatosis and Li-Fraumeni syndrome. There is suggestive evidence that pesticides or toxins in the environment may be associated with glioblastoma, but the data are not strong. Previous radiation therapy given to children for the treatment of tumors can predispose them to glioblastoma. There is controversial data that viruses like cytomegalovirus might play a role in the disease. Despite news reports, there is no evidence that cell phones cause glioblastoma.

E. Antonio ("Nino") Chiocca, chairman of neurosurgery at the Brigham and Women's Hospital, calls the best theory for the origins of glioblastoma "the bad luck gene theory." This theory was published by Bert Vogelstein and his collaborators at Johns Hopkins in 2017. The bad luck gene theory states that one has stem cells in the brain that divide at a certain rate. Each time they divide, there's a chance of an error occurring in a gene that is associated with cellular growth. When an error occurs, the cell grows a bit faster, and because it grows faster, it divides a bit faster. And as the error rate increases, you develop a glioblastoma. Glioblastoma is cell division gone awry that amplifies over time and results in the loss of the normal regulation of cell growth. Unlike smoking and environmental toxins that damage cells and lead to other cancers, there is no evidence that this occurs in glioblastoma. Interestingly, glioblastomas don't metastasize or spread to other areas of the body. They are primarily localized in the brain.

TREATING GLIOBLASTOMA

There have been spectacular advances in the field of cancer therapy based on targeting the immune system, but these have not yet been translated successfully to glioblastoma. The standard of care for the treatment of glioblastoma was defined in 2005 by a landmark study in the *New England Journal of Medicine* and hasn't significantly changed since that time.

The study was headed by Roger Stupp, who asked whether treatment with the chemotherapy agent temozolomide, when added to surgery and radiation, could improve survival in patients with glioblastoma. Temozolomide had shown antitumor activities as a single agent in the treatment of recurrent lower-grade brain tumors. The study was carried out by European and Canadian investigators, with patients receiving temozolomide over a six-month dosing period.

Patients with newly diagnosed microscopically confirmed glioblastoma were randomly assigned to receive radiotherapy alone, five days a week for six weeks, or radiotherapy plus daily doses of temozolomide followed by monthly cycles of temozolomide for six months. The primary goal was overall survival. A total of 573 patients from eighty-five centers were randomized. The median age of participants was fifty-six. Most of the patients had undergone tumor removal surgery. After twenty-eight months of follow-up, the survival was 14.6 months for those who received radiotherapy plus temozolomide, compared to 12.1 months for those who received radiotherapy alone. Adding temozolomide gave patients an increase of two-and-a-half months. The two-year survival rate was 26.5 percent with those given radiotherapy plus temozolomide, compared to 10.4 percent with those given radiotherapy alone. It was a landmark study and was welcomed by the medical community, as the temozolomide was well tolerated. However, improvement by only two-and-a-half months of survival indicated how malignant the tumor was and how hard it was to treat.

Other studies have confirmed the benefit of temozolomide. In 2013, a Japanese analysis of forty-seven patients treated over a five-year period showed survival of 15.8 months in those receiving temozolomide versus twelve months in a control group. Interestingly, the extent of surgery had the strongest impact on survival, arguing for as much removal of the tumor as possible prior to treatment. Furthermore, a major study published in 2017, focusing on older patients, with an average age of seventy-three, showed adding temozolomide to a short-course of radiotherapy resulted in longer survival than a short-course of radiotherapy alone.

However, almost fifteen years after the landmark Stupp study, despite great effort and many trials, the use of temozolomide as standard of care has not changed, although there are people who are surviving longer. I spoke with David Reardon, a medical oncologist at the Dana-Farber Cancer Institute, who has spent his life treating people with glioblastoma. I asked him about the longest period of survival he had ever seen. He told me he cares for a man who has survived eighteen years, whose glioblastoma began at the age of thirty. These rare cases provide a glimmer of hope, although they are few and far between. When I asked David why he thought that patient survived, he said he didn't really know but the patient's youth may have been a factor.

Stupp's team also established the first predictive marker for glioblastoma, the presence of the *MGMT* gene in the tumor. The promoter of the *MGMT* gene is the spot where gene activation begins. It can be altered by a chemical reaction called methylation. Glioblastoma patients with methylation of the *MGMT* promoter do better with temozolomide than patients without the methylation. In a trial of older glioblastoma patients, temozolomide was actually detrimental to those with an unmethylated MGMT promoter in a head-to-head comparison versus radiotherapy. In this same fragile patient population, those with MGMT methylation did best with temozolomide, even without radiotherapy. These studies have led Stupp to conclude that temozolomide and similar chemotherapy drugs are "of marginal benefit, if any" for patients with an MGMT unmethylated glioblastoma. This patient group clearly needs better treatments.

The crime scene for glioblastoma tumor cells—namely, the brain—is unique. The brain's function depends on billions of neurons. Although glioblastoma cells arise from non-neuronal cells in the brain called glial cells, it is now known that activity of the neurons promotes the growth of glial cells in mice and that the neurons secrete a protein called N1GN3, which promotes growth and proliferation. To determine whether this protein was required or could affect tumor growth, investigators took glioblastoma from human subjects and implanted them into mice with a deficient immune system so the mice do not reject human tissue. Remarkably, the transplanted glioblastoma grew in the brains of mice that had the N1GN3 protein, but not in the brains of mice that were genetically deficient in N1GN3. However, breast cancer tumors transplanted into the brain were not affected by a deficiency of N1GN3. The neurons in the brain affected the growth of a brain tumor—glioblastoma—by a brain-specific mechanism that could not affect just any type of tumor put into the brain, such as breast cancer. There was something unique about the brain and the tumors that could grow there. How could this then be turned into a therapy? Initial studies in animals suggest that blocking the production of N1GN3 may slow tumor growth. Now it must be translated into people.

These studies show that the crime scene in glioblastoma is not only composed of the tumor itself but of non-tumor cells as well, such as infiltrating tumor-associated macrophages, described above, and factors released by neurons such as N1GN3. Better understanding the crime scene and ways to affect it opens a pathway to new therapies. Although these animal studies are a long way from clinical trials, this is the only way new therapies for glioblastoma will be developed.

In order to discover new ways to stop tumor growth, it is always good to begin with the tumor itself. Following this principle, investigators took tumors from patients with glioblastoma and placed them in the tissue culture dish and in the mouse brain. They then used RNAi inhibitors, special molecules that inhibit specific RNAs, to determine which of the inhibitors affected tumor growth. Surprisingly, what was found in the tissue culture dish was not always found in the mouse brain. In fact, of the more than 1,500 inhibitory RNAs tested, only three had an effect both in the test tube and in the brain. The lead hit was called JMJD6, which affects the generation of proteins. Targeting JMJD6 extended the survival of mice receiving glioblastoma tumors from patients. These results by the Cleveland Clinic's Tyler E. Miller showed that it may be better to screen tumors by putting them into animal brains, rather than only in the test tube.

Another study, headed by Stanford's Elizabeth Qin, investigated how brain tumors spread. One of the critical areas of the brain is just underneath the ventricles, a region called the subventricular zone. Ventricles are cavities in the brain where the spinal fluid resides. Tumor cells invade the subventricular zone, but it is not known how. Qin and her team measured cells in the subventricular zone and found that a compound called pleiotrophin is highly enriched in those cells, both in mice and humans. Furthermore, these cells send out signals to the tumor cells and tell them how to invade. So, one way to prevent cancer cell invasion is to block compounds that attract tumor cells so they don't invade.

Of course, all of these unique approaches depend on discovering a compound that gets into the brain and does not have unacceptable side effects. When he was at The Ohio State University, Sean Lawler (now at our hospital) discovered compounds called indirubins that come from a Chinese herb. They blocked glioma cells from invading and easily crossed the blood-brain barrier to enter the brain in mouse models. Because they have been known for a long time, indirubins could not be patented, which, unfortunately, reduces interest from pharmaceutical companies.

It is known that human glioblastoma cells harbor a subpopulation of stem cells that drive the tumor, but where these stem cells come from and how they rear their ugly heads is not clear. In a sophisticated approach, Xiaoyang Lan marked individual glioblastoma cells and studied how the cells and their clones grew. Using this individual cell-tracking approach gave researchers at The Hospital for Sick Children in Toronto and the University of Cambridge a real-time look at glioblastoma. The researchers identified two distinct patterns of growth. The first, more common group, followed a typical stem cell hierarchy, where only a minority of stem cells within a clone determined the growth of the clone. The second group consisted of more rare and more aggressive clones. The findings suggested that successful therapy would need to target both groups. From these studies, it appears that human glioblastoma may not come from an aberrant clone but from a normal clone in which the normal developmental program is activated in the wrong way.

Precisely how glioblastomas recur following chemotherapy is unknown. Jian Chen and a team at the University of Texas Southwestern Medical Center directly studied this issue and found that only a small-cell population propagated glioblastoma growth after chemotherapy. Using a mouse model of glioma, they found that a subset of tumor cells was the source of new tumor cells after chemotherapy with temozolomide was given. Subventricular zone neural cells were identified as the guilty cells causing new tumor growth after chemotherapy. They found that if they ablated these cells using an antiviral medication called ganciclovir, growth did not occur when combined with temozolomide and thus tumor development was impeded. The cells they identified were similar to cells that are proposed to be cancer stem cells and may be responsible for long-term tumor growth. It is likely that we need to understand and block the mechanisms by which tumor cells escape regulatory control and grow again, a process that leads to tumor recurrence after surgery, radiation, and chemotherapy.

SURGICAL TREATMENT

Treating glioblastoma always begins with brain surgery. With modern surgical techniques, a glioblastoma tumor can appear to be completely removed, with the MRI showing a clean crime scene with no evidence of tumor. However, the tumor invariably returns because tumor cells escape the edge of the tumor mass and infiltrate normal brain tissue. These microscopic cells cannot be removed by surgery. Nonetheless, the more completely one can remove the tumor, the better chance of the patient surviving longer.

Nino Chiocca has spent his life operating on people with glioblastomas. He's told me how difficult it is to operate on such patients, knowing that the tumor will always return, although each patient responds differently. He told me the intriguing story of a seventy-five-year-old woman who lived for five years with glioblastoma. Each time he operated on her, the tumor recurred a year later. Nino said that he operated on her five times and that those five years allowed her to see her grandchildren graduate from college, and to have great moments with her husband.

Brain surgery has changed dramatically over the twenty-five years since Nino began operating. One of the advances in brain surgery is image guidance. MRI images and sophisticated software give surgeons an almost GPS-like system to guide them to the precise location of the tumor. Before these guidance systems, the surgeon would spend the first half hour just trying to find the tumor. In addition, surgeons can now use intraoperative

MRI or ultrasound, so they can see in real time what's happening as the tumor is being removed. Some surgeons use special exoscopes, allowing them to look at the tumor through a camera lens and operate looking at a screen. Chiocca told me it is important to directly visualize the tumor during surgery in order to minimize bleeding, as that is always a risk given the vascularity of the tumor.

Amazingly, operating on a brain tumor can be relatively straightforward, and the patient usually goes home in two or three days. Many times, because the brain itself does not feel pain, surgery is not a painful procedure and the patient is often awake during surgery. If the surgery is done while the patient is awake and there is no general anesthesia, the patient can leave the hospital the next morning. Some have also experimented with doing operations on this horrible brain tumor as outpatient surgery, with surgery in the morning and the patient going home in the afternoon, although this is not the norm.

I recently took my two sons and grandson to a World Series game at Dodger Stadium, where the Boston Red Sox faced off against the Los Angeles Dodgers. Being from Boston, we were dressed in Red Sox paraphernalia, surrounded by fans dressed in Dodger blue. All the baseball rituals were in full display—the singing of "Take Me Out to the Ball Game," the cheering and high fives by Red Sox and Dodger fans, depending on which team scored a run.

But of all the rituals, one caught me by surprise. At the top of the fifth inning, both teams took to the field and partisan baseball disappeared as the fans stood up and players from both teams held up a card that read STAND UP TO CANCER. Some players held a card with the words MY MOTHER or MY BROTHER. I must admit, tears came to my eyes as I took in the scene and thought of my father, whom I lost to prostate cancer. I thought back to Nixon signing the National Cancer Act of 1971 and declaring a "war on cancer." Then I thought of the announcement only a few weeks earlier that the Nobel Prize in Medicine was awarded for a breakthrough in the treatment of cancer that acted by harnessing the immune system. Nobel Prizes are not given lightly. Cancer is a huge challenge.

NOBEL PRIZE FOR CANCER TREATMENT

The Nobel Prize in Medicine was shared between James Allison of the University of Texas and Tasuku Honjo of Kyoto University in Japan for their discovery of a new way to treat cancer. It began with basic research in immunology, far from any investigation of cancer. In the late 1980s, Allison became interested in how T cells, the immune system's killer cells, work. T cells are triggered by a very complex process involving stimulatory molecules that can both upregulate and downregulate the way the T cells respond. Virtually all biologic processes are regulated. In simple terms, body temperature is regulated by sweating when we are hot and shivering when we are cold. Allison discovered receptors on the surface of T cells that serve as "on" and "off" switches. The switches are called checkpoints.

Jim Allison showed that a molecule called CTLA-4 inhibited the ability of T cells to become activated. Thus, mice deficient in CTLA-4 had severe inflammatory disease marked by unchecked activation of T cells. In other words, CTLA-4 acted as a natural brake on the response of T cells. Cancer therapies up until that time were primarily aimed at attacking the cancer cell by cutting it out or destroying it with chemotherapy or radiation. Allison postulated that the immune system itself was poised to fight cancer but was not strong enough. Furthermore, as we now know, the cancer itself has complex machinery to prevent the immune system from attacking it. Thus, the experiment was simple: remove one of the immune checkpoints and see if an immune system in which one of the brakes had been taken off would fight cancer better. Allison injected animals with an antibody against CTLA-4, then implanted tumors to see if tumor growth was inhibited. The experiment worked!

The anti-CTLA-4 antibody also worked against cancer if there was an established tumor. What was remarkable was that the treatment worked against multiple types of tumors. It appeared that the immune system was ready to fight against cancer if only it could be released to do so. It is now known that the ability of the immune system to attack tumors exists throughout the body as long as it can be freed to attack the tumor. The immune system is constantly protecting us and serves two roles: fighting off infection and battling cancer cells.

The idea of using the immune system to fight cancer is not new. In the nineteenth century, a surgeon named William Coley had a patient who had a cancer called a sarcoma on his face. Coley discovered that the cancer disappeared when the patient also had a severe streptococcal infection on his face that caused erysipelas, a bacterial infection of the skin. This led Coley to develop what became known as "Coley's toxin," in which he injected bacterial products from *Streptococcus* and other bacteria into people with cancer. Although one of his hypotheses was that the bacteria itself had an effect on the cancer, many now believe that the infection was triggering the immune system to fight the tumor. Others had observed this process before. Anton Chekhov, the playwright and physician, observed a relationship between erysipelas and remission of cancer.

Studies in animals do not necessarily translate into an effective treatment in people. Vijay Kuchroo, an immunologist in our center with extensive expertise in the workings of the immune system, won the Dystel Prize for MS Research and the Paul Prize for Cytokine Research. He remembers speaking to Allison when Allison was met with skepticism after presenting his animal data at a prestigious scientific meeting. After Allison's presentation, someone said, "You know, it's well and good, we cure tumors in mice all the time. It will not amount to anything in humans." Alison himself had successfully survived prostate cancer, and he was undaunted in wanting to test anti-CTLA-4 in people with cancer. He said, "We will try it in cancer patients and let the data speak for itself." He had to work hard to find funding to get anti-CTLA-4 antibody to test in people. With the help of the National Cancer Institute, he moved forward.

The clinical investigation of anti-CTLA-4 was done in patients with advanced melanoma. Although not all patients responded positively, some had dramatic reactions and appeared to be cured. Interestingly, it often took time for the effect to be observed, as the immune system first had to be mobilized. Ipilimumab, a monoclonal anti-CTLA-4 antibody, was approved by the FDA in 2011 for the treatment of inoperable or metastatic melanoma. CTLA-4 is not the only brake or inhibitor on T cells, and investigators studied other checkpoint inhibitors, including one called PD-1 that was pioneered by Tasuku Honjo, who won the Nobel Prize with Allison. In 2014, the FDA approved nivolumab, a monoclonal antibody against PD-1.

Checkpoint inhibitors have revolutionized the treatment of cancer and have launched the field of immuno-oncology. Many new checkpoint inhibitors are being studied. Nivolumab is widely used in cancer, including melanoma, non-small cell lung cancer, renal cell carcinoma, Hodgkin's lymphoma, head and neck cancer, urethral carcinoma, and hepatocellular carcinoma. Ipilimumab and nivolumab are often given in combination, and testing new approaches and drugs in cancer includes the use of these checkpoint inhibitors.

Former president Jimmy Carter had metastatic melanoma and was treated with pembrolizumab, an anti-PD-1 checkpoint inhibitor, which he received at the age of ninety-one. His tumor disappeared. The use of checkpoint inhibitors, however, is not without side effects. Activating the immune system can cause problems with autoimmunity, which happens with MS. Taking the brakes off the immune system with a checkpoint inhibitor allows the immune system to attack not only the cancer cells but also normal cells of the body. Side effects of checkpoint inhibitors are the induction of autoimmune syndromes, including colitis, rashes, and, in some cases, myasthenia gravis, an autoimmune disease.

With all the excitement in the cancer field surrounding checkpoint inhibitors, the critical question became, How effective would they be against glioblastoma? Were they the long-awaited breakthrough? A major test of the checkpoint inhibitor nivolumab was conducted by Bristol Myers Squibb in a trial called CheckMate 143. The randomized Phase 3 trial was carried out in patients with the first recurrence of glioblastoma. In April 2017, Bristol Myers Squibb announced the results. Despite a great deal of both hype and hope, the trial was negative and did not improve overall survival. Nonetheless, additional trials of checkpoint inhibitors are underway, one called CheckMate 498 and the other CheckMate 548, in which a PD-1 checkpoint inhibitor will be combined with radiation therapy with or without chemotherapy.

Despite the success of anti-CTLA-4 and anti-PD-1, not all new checkpoint inhibitors work. To the surprise of many, a new checkpoint inhibitor targeted to the IDO1 protein worked well in Phase 2 trials but failed in Phase 3 trials; as we have seen, Phase 2 trials do not always predict how a drug will work in a Phase 3 trial. The field will need to reevaluate whether the problem was with the drug or the target. Some believe that all the excitement surrounding checkpoint inhibitors might have resulted in the IDO trials moving too quickly.

Although checkpoint inhibitors clearly helped Jimmy Carter, they couldn't help John McCain. Glioblastoma is not the only tumor that does not respond to checkpoint inhibitors. Most colorectal cancers are known for not responding to checkpoint inhibitors. Why checkpoint inhibitors would not work in glioblastoma relates to the complex nature of glioblastoma and the strong defenses against the immune system that exist at the crime scene. Careful examination of the crime scene in glioblastoma shows many features that make it difficult not only for checkpoint inhibitors to work but for other forms of immune therapy to work as well.

One of the unique features of glioblastoma is that it is located in the central nervous system, which is considered an immune-privileged site. In other words, it is protected from the immune system. More than fifty years ago, Peter Medawar showed that foreign cells implanted into the brains of rodents became successfully engrafted, whereas the same cells were eradicated by the immune system when they were placed outside the brain.

Theoretically, if you transplanted a small piece of foreign kidney in the brain, the immune system would not reject it, whereas if a foreign kidney were implanted anywhere else in the body, there would be kidney graft rejection. Another organ that has immune privileged features is the gut, as the body does not want over-reactive immune responses in the gut where food is absorbed. This could be why checkpoint inhibitors don't work well against brain tumors or colorectal cancer.

Many other features of glioblastoma protect it from attack by the immune system. Glioblastomas are loaded with TGF-beta, a very suppressive immune molecule. Additionally, T cells that enter the brains of patients with glioblastoma are also loaded with suppressive molecules that not only include TGF-beta but many other suppressive molecules as well. All of these molecules put a brake on the immune system. Microglial cells in the brains of patients with glioblastoma have tons of the suppressive chemicals TGF-beta and IL-10. Cytotoxic, or killer, cells are crucial for destroying tumors in glioblastoma, but because of factors produced by the tumor, the cytotoxic cells are dysfunctional and unable to kill. Not only is there suppression of the immune system in the brain but the glioblastoma also suppresses the immune system throughout the body, making patients more susceptible to infections. Glioblastomas put up an amazing defense against the attacking immune system.

Paradoxically, treatment that is given for glioblastoma can decrease the activity of the immune system. Radiation therapy as well as temozolomide, the chemotherapy drug used to treat glioblastoma, cause a decrease in T cells and may cause destruction of tumor-specific killer T cells in the spleen. There's a paucity of T cells in the glioblastoma tumor microenvironment, which is in striking contrast to other tumors, such as melanoma and lung cancer. Instead, there are many myeloid-type white blood cells that suppress the immune system's ability to attack the cancer, and it is possible that these myeloid cells could be a target for future glioblastoma immunotherapy.

Another important concept related to glioblastoma is that of cold versus hot tumors, a distinction that is defined by the extent to which the tumor is infiltrated by immune cells. A cold tumor has few infiltrating immune cells, and they are sparsely scattered in the tumor. A hot tumor has many tumor-infiltrating immune cells. Given the negative results of the CheckMate 143 trial of the checkpoint inhibitor nivolumab, many believe that glioblastomas are cold tumors. Another feature of glioblastoma that makes it a cold tumor is necrosis, or cell death, which makes it more difficult for immune cells to infiltrate the tumor. Thus, glioblastoma is considered to be one of the most immune cold tumors of all cancers, and this contributes to the difficulty of treating it with immune therapy.

Finally, glioblastomas are considered to have one of the lower rates of tumor mutations relative to other cancers. Each tumor mutation creates a potential new target for the immune system to attack. Thus, a tumor with only a few mutations provides fewer targets for the immune system to attack, which in turn leads to a poor response to checkpoint inhibitors.

One possibility for overcoming the failure of checkpoint inhibitors in the CheckMate 143 trial relates to the fact that the chemotherapy drug temozolomide suppresses the immune system. Perhaps newly diagnosed patients would have a better response to checkpoint inhibitors if they weren't given chemotherapy. With this in mind, Bristol Myers Squibb has two ongoing trials of a checkpoint inhibitor named Opdivo, or anti-PD1, which is given with radiotherapy, both with and without temozolomide in patients with newly diagnosed glioblastoma. Another strategy could be to increase T cell infiltration in the tumor by combining checkpoint inhibitors with vaccines to activate cold tumors. If T cells against the tumor were induced by immunizing with tumor proteins, this could force these T cells into the tumor, leading to warmer tumors and a better response to checkpoint inhibitors.

VIRUSES TO TREAT GLIOBLASTOMA

If checkpoint inhibitors given by intravenous infusion did not work in glioblastoma, perhaps a bolder approach, such as directly activating the immune system in the brain, was required. The immune system evolved to fight infection, so what if a virus was injected directly into the tumor? The injected virus would serve as a strong natural signal to attract immune cells to the virus inside the tumor. When the immune cells entered the tumor to kill the virus, they would also kill the tumor. This approach is called oncolytic virus therapy.

An oncolytic virus is a virus that infects and kills cancer cells. Its use was first reported in 1911 by an Italian gynecologist who injected a cervical cancer with a rabies vaccine and showed that the cancer disappeared. Over the years, several crude preparations of viruses were injected in patients with cancer. However, it was not until the early 1990s that Robert Martuza at Massachusetts General Hospital engineered a tumor-specific virus based on herpes simplex virus type 1 and used it in the first clinical trial in humans with glioblastoma. Shortly thereafter, Nino Chiocca began to inject a different oncolytic virus, based on the cold virus (adenovirus) in a clinical trial. There were twenty-four patients in the clinical trial, and two are still alive more than fifteen years after being injected! The major lesson from these trials was that oncolytic viruses could be safely injected in the brain without causing bad viral infections in patients.

Following the pioneering work of Martuza and Chiocca, another approach used poliovirus, one of the classic viruses in medical history. The road to treating glioblastoma by injecting poliovirus into the brain began in 1996 with basic work in the laboratory of Matthias Gromeier, now a professor in Duke's Department of Neurosurgery. More than twenty years later, poliovirus-directed therapy of glioblastoma was featured on *60 Minutes* and on *CBS Newsmagazine.* Results were published in the *New England Journal of Medicine* in July 2018.

If a virus is going to be injected into the brain to help treat glioblastoma, it can't be infectious. So, the first step is to modify the virus so it is not infectious but can still stimulate the immune system. In 1996, Gromeier published an article about his modification of the poliovirus so it was no longer toxic to nerve cells. Then in 2000, after testing the virus on mice with experimentally induced gliomas, he reported on the strong oncolytic, or tumor-killing, potential of modified poliovirus for the treatment of malignant glioma. One of the key reasons for choosing poliovirus is that the receptor for the virus, called CD155, is found on glial tumors, such as glioblastoma. Indeed, the location of CD155, both in the gut and on motor neurons of the spinal cord and brain stem, is why polio causes paralysis.

Poliovirus infects only humans and primates. To prove that CD155 was an important receptor, Gromeier carried out experiments in which mice that had CD155 in their tissues developed a polio-like syndrome after poliovirus infection. Gromeier and his colleagues then found that the modified poliovirus worked best on glioblastoma when it was injected directly into the tumor rather than given intravenously, and thus human trials evolved that entailed a direct injection of the virus into the tumor. Of course, a virus will work only if the tumor cells express CD155, and luckily the majority of glioblastomas have CD155 on their surface.

As it turns out, CD155 is on other tumors as well, and the recombinant poliovirus is now also being considered for treating other tumors. In our early work on ALS, Bob Brown and I studied whether poliovirus could provide a clue that would lead to a treatment for ALS. At that time, we didn't know that CD155 was the receptor for poliovirus. Although an interesting hypothesis, it did not help us find a treatment for ALS or connect poliovirus to ALS.

Based on Gromeier's work, a genetically constructed recombinant poliovirus called PVSRIPO was manufactured in association with the Biopharmaceutical Development Program at the National Cancer Institute in Frederick, Maryland, for administration to patients with glioblastoma. It was supported both by philanthropy and a start-up company called Istari Oncology. Because seven of the authors on the *New England Journal of Medicine* paper had equity in the Istari Oncology start-up company, an external Data Safety and Monitoring Board oversaw the conduct of the Duke study.

The Duke study enrolled sixty-one adult patients who had recurrent glioblastoma confirmed by histological testing. The study evaluated seven doses of PVSRIPO. The patients were enrolled over a five-year period between May 2012 and May 2017 and were compared with a historical control group that had been treated at Duke University Medical Center according to standard of care. Toxicity was the primary outcome measure of this Phase 1 trial, with overall survival as a secondary outcome measure.

Over the course of the study, the treatment was granted breakthrough status by the FDA. Breakthrough status came into law in 2012 in response to drugs being developed for specific categories of patients who have serious or life-threatening disease. For breakthrough status to be awarded, there must be favorable preliminary data from clinical trials in humans. During the five years when this trial was being performed, the FDA approved forty-six drugs with breakthrough therapy designation on the basis of eighty-nine pivotal trials. Half were cancer drugs.

By the time the *New England Journal of Medicine* published results of the trial in July 2018 in an article entitled "Recurrent Glioblastoma Treated with Recombinant Poliovirus," the five-year trial had already been featured both on *CBS Newsmagazine* and *60 Minutes. CBS Newsmagazine* first aired a segment in 2015, featuring Stephanie Hopper, a nursing student diagnosed with a late-stage glioblastoma at age twenty. After standard treatment with surgery, chemotherapy, and radiation, her tumor came back. In 2012 she became the first person to be treated with the recombinant PVS-RIPO poliovirus. Stephanie was fully conscious during the six-and-a-half-hour brain surgery when a tiny hole was drilled into her skull and a catheter inserted into her brain. The salt solution containing the virus was delivered to her frontal lobe. After her surgery, there was swelling in her brain, which could have been a sign the tumor was growing. Instead, the swelling was a positive response, because her immune system was activated to fight the tumor. Follow-up scans showed the disappearance of the tumor, and at the time the *60 Minutes* segment was produced, Stephanie had survived more than two years since her surgery. Six years later she was still cancer free, working as an oncology nurse. The program also featured a second patient, a retired cardiologist, who, at the age of seventy, was the second person in the polio trial. He was doing well after three years.

Nothing could have been more dramatic than the *60 Minutes* report. I have been involved in my share of the media spotlight on research and clinical trials, beginning with our report in the *New England Journal of Medicine* about chemotherapy for the treatment of MS. As a writer and filmmaker, I'm acutely aware of the importance of narrative. When it comes to disease, the narrative everyone is searching for is that of a miracle cure. The news of something failing, or something merely marginal, is not compelling, although those incremental steps are crucial on the road to finding a cure.

On the *60 Minutes* program, we heard Scott Pelley report breathlessly, "In just a moment, polio will be dripped into the brain of fifty-eight-year-old Nancy Justice. Her glioblastoma tumor was discovered in 2012. Surgery, chemotherapy, and radiation bought her two and a half years. But the tumor came roaring back. Now, the poliovirus in this syringe, which mankind has fought to eradicate from the earth, is the last chance she has in the world."

Sadly, despite her courage and hope, Nancy suffered a recurrence of her glioblastoma and died on April 6 at the age of sixty. Nancy was the seventeenth patient treated by the Duke team. The next stage in investigating the poliovirus approach was to use a higher dose of the virus. It is standard to search for the maximal tolerated dose of a therapy with the belief that the more that can be given safely, the better. On the same program, the show featured a woman named Donna Clegg, who received a higher dose, but sadly the immune reaction with the higher dose was too much for her. She became partially paralyzed and subsequently died. She had dropped out of the trial, and when asked whether she would have done it again, she'd replied, "I may not have done it."

In my experience, when someone enters an experimental trial, no matter how long the odds, they secretly believe they will be the lucky one. It is human nature. Although there was the requisite hype needed for any popular medical reporting, CBS was certainly justified in highlighting this exciting experiment, which goes to the heart of the complex mechanisms required to treat glioblastoma and cancer itself.

When the results were finally published, of the sixty-one poliovirus-treated patients, the median overall survival was 12.5 months compared to 11.3 months for the historical control group. Not a particularly dramatic difference. However, two years after the treatment, the survival curves began to diverge. At twenty-four months, it was 21 percent survival in those treated versus 14 percent in the control group, and at three years, survival was 21 percent of the treated patients compared to 4 percent of the control group. This long-tail subset of survivors suggests that some patients might in fact have been cured, although the survival benefit might have been related to patients with good prognostic characteristics. In addition, almost half of the patients received a drug called bevacizumab, which dampens blood vessel growth to treat the symptomatic swelling around the tumor, and this might have confounded the results. Nonetheless, a three-year survival rate of 21 percent is impressive. The treatment was not an easy one, and 19 percent of the patients experienced severe side effects from therapy, including brain swelling, seizures, and muscle weakness.

In the almost twenty years since the recombinant poliovirus was first made, a great deal more is known about its mechanism of action. The sequence of events is as follows: First, the live-attenuated poliovirus binds to human cancer cells via the CD155 viral receptor. It kills cancer cells, which in turn releases cancer proteins, which themselves activate the immune system. Second, the virus infects dendritic cells of the immune system and makes them more stimulatory, while at the same time it reactivates the paralyzed dendritic cells and macrophages that the tumor stopped to prevent the immune system from attacking. Finally, the poliovirus-activated dendritic cells induce killer T cells, which then further attack the tumor and prevent its recurrence. A bonus is that additional white blood cells called neutrophils are recruited following dendritic cell and T cell stimulation. This complex viral and immune cascade fights the tumor.

Remember, the immune system evolved to deal with infections, so it is not surprising that appropriately designed infection could trigger the immune system to fight cancer.

The recombinant poliovirus PVSRIPO was used as part of a strategy that employs oncolytic viruses programmed to infect and kill tumor cells. Although this tumor-killing effect was the initial rationale behind oncolytic viruses, it is now recognized that it may very well be the stimulation of the immune system by multiple mechanisms after viral infection that has the ultimate impact. It has become a cornerstone of cancer immunotherapy that the immune system must be stimulated, and the clinical development of oncolytic viruses is more and more focused on their ability to stimulate the immune system rather than simply destroy cancer cells. Currently, there are more than forty clinical trials recruiting patients for treatment with oncolytic viruses, including more than fifteen trials for glioblastoma alone. In addition to poliovirus, other viruses being tested include measles virus, herpes virus, and adenovirus. There are also clinical trials that combine an oncolytic virus to stimulate T cells to infiltrate the glioblastoma, plus an immune checkpoint inhibitor to activate these T cells to attack the glioblastoma.

With more than fifteen oncolytic virus trials underway in glioblastoma, this could be the first sign of progress against the disease. The next steps, which involve combination therapy with both oncolytic viruses and immunotherapy, are moving forward. Imlygic is the first FDA-approved oncolytic viral therapy in the United States. It was approved in 2015 for use in advanced melanoma. Imlygic is a genetically modified herpes simplex virus designed to replicate in the tumor. The cost of therapy is approximately $65,000 per year.

Despite the attention brought to poliovirus by the *60 Minutes* story, the oncolytic virus therapy farthest along for glioblastoma is called Toca 511. Toca 511 is derived from a Moloney murine leukemia virus and modified to preferentially infect tumor cells. During surgery, the virus is injected into the tissues lining the cavity where the tumor was removed. Toca 511 doesn't kill the tumor cells. Instead, it is designed to spread and infect the remaining cancer cells with a gene that causes the cancer cells to produce a protein called cytosine deaminase (CD). Six weeks later, a drug is injected to activate CD against the cancer. It is a two-step therapy designed to weaponize the tumor against itself. Unfortunately, although there were positive effects in Phase 2 studies, a Phase 3 pivotal, randomized, double-blind study did not succeed.

With the idea of treating early, Nino Chiocca published a recent Phase 2 study in which he used a virus when the glioblastoma was first diagnosed, not when it recurred. When patients are first diagnosed, they have a much stronger immune system. In addition, this approach allows the patient to get the virus plus standard of care treatment with radiation and chemotherapy. This therapy showed that over a third of the patients were still alive at three years after treatment, compared to only 6 percent of patients who only had standard of care. A Phase 3 trial is being planned.

Oncolytic viruses provide a new tool for combating glioblastoma, and they appear to work by enhancing the immune response against the tumor. This is done both by killing tumor cells and releasing tumor proteins, and by directly overcoming tumor-induced suppression of the immune response. The ultimate success of oncolytic viruses will depend on identifying which populations respond best to them, judicious use of drugs so they don't interfere with the ability of the immune system to destroy the tumor, and combination therapy. DNAtrix is a biotechnology company testing combination therapy of an oncolytic virus with a checkpoint inhibitor in glioblastoma.

CANCER AND THE GUT

Although the gut is far away from the brain, manipulating the gut can also affect other diseases of the nervous system, including Alzheimer's, ALS, and Parkinson's disease, suggesting that there is a connection between the gut and the brain in many neurological diseases, as we have discussed. This connection has been most studied in MS, which is logical because this disease is mediated by the immune system, and we know the gut is very important in driving the immune response. Remarkably, the gut has now been shown to play an important role in cancer therapy and has direct relevance to glioblastoma. Indeed, there are drug companies that are attempting to develop therapeutics for glioblastoma based on the microbiome.

In the 1980s we used the chemotherapy drug cyclophosphamide to treat MS. It has now been shown that the anticancer effects of cyclophosphamide are dependent on the gut. The studies on cyclophosphamide and the microbiome in cancer were first published in 2013 by Laurence Zitvogel's group in Paris. They discovered cyclophosphamide lost its antitumor effects in an animal model of cancer if they eliminated the microbiome (bacteria in the gut) by using antibiotics.

The microbiome works to help the anticancer response because activating the immune system can help destroy cancer cells, and the cyclophosphamide treatment resulted in increased numbers of killer T cells of the Th17 type in the spleens of animals. The mechanism went as follows: cyclophosphamide treatment disrupted the gut barrier and allowed the release of bacteria, which then traveled to the spleen, where they stimulated the immune system. Thus, the gut served as a source of immune stimulation. We know that one of the barriers to immune therapy of glioblastomas and other types of cancer is a strong immunosuppressive environment. Stimulating the immune system via the gut can overcome that.

Zitvogel's team then identified which type of gut bacteria was responsible for the anticancer effect of cyclophosphamide by treating animals with different types of antibiotics. There are two general classes of bacteria, gram-positive and gram-negative. One antibiotic removed gram-positive bacteria from the gut, while the other removed gram-negative bacteria. It was the gram-positive bacteria that were responsible for the anticancer effect of cyclophosphamide. The gram-positive bacteria served as an adjuvant and increased Th17 killer cells. If the gram-positive bacteria were removed from the gut, the cyclophosphamide had no antitumor effects. Furthermore, injecting Th17 cells into animals could restore the antitumor efficacy of cyclophosphamide. Three years later, Zitvogel's group identified the specific bacteria responsible for this effect in animals and showed that it was important for humans with cancer. The gram-positive bacteria responsible were called *Enterococcus hirae* and *Barnesiella intestinihominis*.

Zitvogel's group then tested patients who received chemotherapy for cancer to determine whether an immune response occurred related to these gram-positive bacteria. The answer was yes. Patients with advanced lung and ovarian cancer whose immune systems had reacted against the bacteria lived longer and had a longer period of time after treatment before the disease began progressing again. The exact mechanism and how this worked are not clear but could relate to the bacteria affecting the overall efficacy of the immune system. More intriguingly, however, is the possibility that structures on the bacteria were similar to structures on the tumor, and releasing these bacteria into the immune system stimulated the immune system to develop antitumor responses, a type of natural antitumor vaccine.

The results with cyclophosphamide were all well and good, but it was important to understand how broadly applicable this observation was. Specifically, how did it relate to the Nobel Prize–winning breakthrough in cancer therapy with checkpoint inhibitors such as CTLA-4 blockade and anti-PD-L1? Not all patients respond to checkpoint inhibitors, so the key question was whether the lack of response could somehow be related to the microbiome. Two studies in mice showed that this was the case.

Zitvogel's group studied anti-CTLA-4, the first checkpoint inhibitor approved for treatment of humans following James Allison's pioneering studies. Anti-CTLA-4 therapy was effective in mice housed in a facility that allowed a normal microbiome, but it was not effective in mice in completely germ-free facilities. The gut microbiome is depleted in germ-free facilities, as germs are needed to stimulate the microbiome. There are two ways to test whether the microbiome is required for a biologic effect: first, to treat the mice with antibiotics to wipe out the bacteria in the gut; and second, to test animals housed in germ-free facilities. The investigators found that both of these maneuvers decreased the efficacy of anti-CTLA-4 therapy.

They then studied how the microbiome was affected by these treatments and found that organisms in the genus *Bacteroides* played an important role in the anti-CTLA-4 effect. *Bacteroides* are the most predominant anaerobic bacteria in the gut. Treatment with the antibiotic vancomycin caused an increase of *Bacteroides* and was associated with a better response to anti-CTLA-4 therapy. Remember, vancomycin had the opposite effect in mice treated with the chemotherapy drug cyclophosphamide, indicating that a bacterium may act differently depending on the type of cancer treatment.

The investigators then performed the most important experiment to prove the point. They took feces from patients with melanoma who harbored *Bacteroides* species and transferred them into the intestines of mice. They found that feces from the cancer patients with increased *Bacteroides* enhanced the effects of anti-CTLA-4 in mice. They also found that immune responses against *Bacteroides* were associated with efficacy of anti-CTLA-4 both in mice and patients. It is fascinating that treatment with anti-CTLA-4 in patients favored the outgrowth of *Bacteroides* with anticancer properties.

The other major anticancer checkpoint inhibitor is anti-PD-L1, which today is in more widespread use in humans than anti-CTLA-4, given that it has fewer side effects. The microbiome is also important in anti-PD-L1 therapy. In a study done by Thomas Gajewski's group at the University of Chicago and published in *Science*, which also published Zitvogel's study, they found that the *Bifidobacterium* bacteria promotes antitumor immunity and facilitates anti-PD-L1 efficacy. Gajewski's group studied the mouse model of melanoma and found that identical mice raised in two different animal facilities—the Jackson Laboratory and Taconic Biosciences—responded differently to anti-PD-L1 therapy. It is known that mice from these two facilities have distinct microbiota and have different tumor growth. The melanoma tumors grew more aggressively in

Taconic mice than in Jackson mice, and this was linked with increased tumor-specific T cell responses and accumulation of CD8 killer cells in the tumor in Jackson mice relative to Taconic mice.

To determine whether the differences between facilities was related to the microbiome, they performed a simple experiment: they put mice from two facilities in the same cage, a practice known as co-housing. If mice are co-housed, they ingest feces from each other, resulting in the natural transplantation of feces. The researchers found that co-housed Taconic mice acquired the characteristics of the Jackson mice, suggesting that the Jackson mice were colonized by bacteria that dominated and facilitated antitumor immunity. The investigators then identified *Bifidobacterium* as the bacteria with the antitumor effects. Administering *Bifidobacterium* alone inhibited tumor growth to the same degree as treating with anti-PD-L1, the checkpoint blockade antibody, and if both were used, tumor growth was nearly eradicated. The *Bifidobacterium* increased the killer activity of the immune system, but the *Bifidobacterium* had to be alive to have its effects.

The stage was now set to determine whether the effects on the microbiome in patients' responses to checkpoint inhibitors could be shown in humans with cancer. The answer was a resounding yes, with the publication in 2018 of three papers in *Science*, in which the gut microbiome was studied in cancer patients treated with anti-PD-1. Laurence Zitvogel reported that the gut microbiome was important in patients treated with anti-PD-1 therapy for epithelial tumors, and teams led by James L. Gajewski and Jennifer Wargo at the University of Texas MD Anderson Cancer Center showed the importance of the gut microbiome in patients with metastatic melanoma who were treated with anti-PD-1.

Zitvogel's group asked whether antibiotics, which wipe out the microbiome, would affect the response of patients with advanced non–small cell lung cancer, renal cell carcinoma, and urethral carcinoma. They analyzed 249 patients who were given antibiotics two months before or one month after the first administration of anti-PD-1 and measured both progression-free survival and overall survival. They found that people who were treated with antibiotics did not live as long. They further tested these findings in a validation cohort of 239 patients with advanced non–small cell lung cancer and confirmed that antibiotic treatment had a negative impact. They then performed fecal microbiota transplantation (FMT) from cancer patients who responded to anti-PD-1 into mice. They found that transplanting fecal material from people who responded to therapy affected tumor growth but the transfer of feces from nonresponding patients failed to do so. When they analyzed the gut contents of the patients, they found an abundance of the bacteria *Akkermansia muciniphila*. Indeed, orally supplementing nonresponder feces with *A. muciniphila* restored the anti-PD-1 effect, proving that *Akkermanisa* was the specific bacteria important for the effects. Furthermore, they found that patients who responded had immune reactivity against *Akkermansia*, one of the most abundant bacteria in the lower intestine.

Gajewski's experiments analyzed stool samples from metastatic melanoma patients before anti-PD-1 treatment and found that there was an association between the microbiome composition and the clinical response to anti-PD-1. They identified three bacteria that were more abundant in the responders, including *Bifidobacterium longum*, *Collinsella aerofaciens*, and *Enterococcus faecium*. When they transferred fecal material into mice from anti-PD-1 responding patients, the mice responded better to anti-PD-L1 therapy and also had enhanced T cell responses. Jennifer Wargo's group also found gut microbiome differences between people who responded and people who didn't respond to anti-PD-1 treatment of melanoma. They identified the responsible bacteria to be in the Ruminococcaceae family.

With these dramatic results, investigators are now applying their findings to the treatment of cancer patients. Zitvogel advises caution about giving antibiotics to patients undergoing checkpoint inhibitor therapy. Wargo is planning to manipulate the gut microbiome with fecal transplants given in a pill form in a study sponsored by the Parker Institute for Cancer Immunotherapy in San Francisco. A group at the University of Pittsburgh has partnered with the global pharmaceutical company Merck to collect fecal bacteria from people who respond to treatment with a checkpoint inhibitor and to transfer it into the intestines of nonresponders. Gajewski has partnered with a biotech company in Cambridge, Massachusetts, called Evelo Biosciences, to test the effect of pills containing a single bacteria strain in people with different types of colon and skin cancer.

These revolutionary findings for treating cancer have direct implications for glioblastoma. A company called Enterome in Paris, France, is pursuing therapy based on the microbiome and has identified glioblastoma as one of its target diseases. Enterome's approach is to identify unique aspects of the microbiome in patients with glioblastoma to determine whether there are specific proteins that react with the glioblastoma. Those that are reactive with the glioblastoma will be used to generate a vaccine.

ORAL TOLERANCE

For myself and for many scientists, the road to discovery is long and often takes unexpected turns. My primary scientific goal for my research in MS has been to find ways to quiet an overactive immune system. I made one of my major discoveries after we published our research showing that suppressing the immune system with the chemotherapy drug cyclophosphamide had a dramatic effect in MS. I then began to look for more natural, physiologic ways to suppress the immune system and turned to the phenomenon of oral tolerance. Oral tolerance is an immune process that prevents the body from mounting an immune response to what we eat. In oral tolerance, the gut naturally tolerates food and other proteins.

The concept of oral tolerance had been known for over a century, but its successful translation for treating autoimmune diseases had not yet occurred. In 1988, I reported that having mice eat a brain protein induced regulatory T cells in the gut, which then traveled to the area of the brain where the brain protein was expressed and suppressed the animal model of MS. I showed a similar treatment effect for other diseases, in which we fed animals collagen to treat arthritis and insulin to treat type 1 diabetes.

It was a very exciting time, with papers published in *Science* and *Nature*. Ultimately we formed a public company called Autoimmune, Inc., with Bob Bishop as CEO, to commercialize oral tolerance.

One of my major discoveries was that oral tolerance worked by introducing a T cell in the gut that suppressed the overactive immune response. The suppressor T cell acted by secreting TGF-beta, one of the suppressive chemicals that we have seen is overabundant in cancer. At that time, the concept of suppressor cells was out of favor in the scientific community. Nonetheless, the journal *Science* published our research identifying suppressor T cell clones that secreted TGF-beta. In those earlier days of T cell immunology there were only Th1 and Th2 cells, and we called our suppressor cells Th3 cells. Today we have Th17 cells. Suppressor cells were ultimately renamed regulatory cells and are now a well-accepted component of the immune system.

In 1995, Autoimmune, Inc. began clinical trials of oral tolerance in MS and in rheumatoid arthritis. The writer Susan Quinn, who had written an acclaimed book on Marie Curie, was intrigued by our work and asked whether she could follow the process and write a book about it. At first I was hesitant, but I ultimately agreed. Why not give the world an inside look at our attempt to translate a basic finding in animals into a treatment for people? What an amazing thing it would be if it worked—a natural, nontoxic way to treat disease.

Sue attended our lab meetings, and I gave her access to my journals. She even interviewed my mother in Denver. Unfortunately, although the Phase 2 trials were positive, to my utter disappointment, all the Phase 3 trials of oral tolerance failed. Autoimmune, Inc. folded. What Susan Quinn had hoped would be a success story turned out to be the story of failure. She entitled her book *Human Trials: Scientists, Investors, and Patients in the Quest for a Cure.* The phrase "human trials" had a double meaning—the actual trials being conducted and the trials those performing the trials had to go through. I will always have great empathy for all the people described in this book, who worked diligently with so much hope to find treatments, only to have their Phase 3 trial fail. With the failure of the oral tolerance trials, my colleagues said to me, "Howard, what are you going to do? All your work is for naught." Like everyone involved in a promising Phase 3 trial that failed, I had no choice but to take the blow. For the next decade I tried to determine how I could make oral tolerance work and continued to investigate the regulatory T cells that were produced in the gut and were responsible for suppressing the immune system and mediating oral tolerance.

A breakthrough occurred a decade later, in 2006, when my lab discovered it wasn't necessary to feed a subject a protein to induce oral tolerance. We found that we could simply feed them a monoclonal antibody called anti-CD3, which triggered T cells in the gut to suppress autoimmune diseases in animals such as EAE, arthritis, and diabetes. Anti-CD3 bound to the T cell receptor and triggered regulatory T cells in the environment of the gut. We also discovered that the regulatory cells we were inducing in the gut had a structure on the surface called LAP, or latency-associated peptide. LAP is part of the cell membrane and contains the suppressive chemical TGF-beta.

No one had ever given a monoclonal antibody orally before; it had always been given intravenously. We now had a path forward to once again use oral tolerance in the clinic, and we expanded this concept to include administering the monoclonal anti-CD3 antibody nasally as well. Oral tolerance was back in business. Over the next decade, I found that oral and nasal administration of anti-CD3 monoclonal antibody ameliorated disease in a large number of animal models, including MS, diabetes, arthritis, and even in Alzheimer's disease. As discussed, plans are underway to initiate human trials by Tiziana Life Sciences. Oral and nasal tolerance with anti-CD3 monoclonal antibody may or may not work, but I have another chance at oral tolerance.

As so often happens, scientific investigation takes unexpected turns, and this is exactly what happened as we continued to investigate the regulatory T cells we discovered in our oral tolerance experiments. Taka Oida, a Japanese postdoctoral fellow, joined the lab around this time with a particular interest in the regulatory T cells we had identified during oral tolerance. He was particularly interested in LAP and wanted to develop an antibody against LAP because there were no good anti-LAP antibodies that could be used to track LAP-positive cells in the mouse. Although there were anti-LAP antibodies for humans, there were none yet for mice. In order to study the basic biology of the immune system, mice would ultimately need to be studied. Taka's idea was to immunize mice with a cell line that expressed LAPs to generate antibodies. Vijay Kuchroo in our center said he didn't think it would work. Taka was relentless, often sleeping in the lab at night to perform his experiments. He refused other people's help because he didn't trust their technique.

Taka succeeded. He made thirty-six monoclonal antibodies against LAP and showed how the expression of LAP was related to FOXP3, a classic marker of regulatory T cells. The work was well done but not conceptually innovative. Thus, we published his paper in 2014 in *PLOS ONE*, a scientific journal that, unlike other journals, does not demand scientific innovation as a criterion for publication, only that experiments are rigorously done. We published Taka's antibodies with the title "TGF-beta Induces Surface LAP Expression on Murine CD4 T Cells Independent of Foxp3 Induction." Under the conclusion and significance section, we wrote that surface LAP on mouse CD4 cells was controlled by FOXP3 and that our newly described anti-mouse-LAP monoclonal antibodies will provide a useful tool for the investigation and functional analysis of T cells that express LAP on their surface.

Given that TGF-beta secreting T cells with LAP on their surface are important for oral tolerance, it was only logical that if an animal was treated with anti-LAP antibodies, it should reverse oral tolerance. This is exactly what happened. We found that the anti-LAP antibody reversed the protective effects of oral tolerance when anti-CD3 was given orally. Again, not a major observation and in a way, expected. We published these results in 2014 in a mid-level immunology journal, *International Immunology*. We reported that mice treated with anti-LAP antibody had a more severe disease in the mouse model of MS. This was a time when people were beginning to investigate checkpoint inhibitors for treating cancer. I then realized that the true value of the anti-LAP antibodies may not have been to understand mechanisms of oral tolerance in MS but as a new checkpoint inhibitor for treating cancer, because anti-LAP antibodies took the brakes off the immune system and could enhance the immune response against tumors.

TGF-beta was known to be very important in cancer. Furthermore, there was no real way to target regulatory T cells such as FOXP3 cells because there was no surface structure by which you could target them. The anti-LAP antibodies we made to study oral tolerance could serve that purpose. We wrote at the end of our 2014 *PLOS ONE* paper that regulatory T cells, or Tregs, not only modulated autoimmunity but could also promote malignancy. Indeed, studies in colorectal cancer showed that LAP+ cells accumulated in tumor sites and were associated with cancer progression. Without mentioning the word *cancer*, we wrote, "Thus, it is possible that anti-LAP antibody administration could have a role in the therapy of diseases where excessive immune regulation contributes to disease processes."

How to conduct the experiments with our unique anti-LAP antibodies was obvious. Take tumor models in mice, implant tumors, and see if treatment with anti-LAP antibody decreased tumor growth. Galina Gabriely, a senior scientist in our lab, undertook the project, and from the very beginning the experiments worked beautifully. She found that anti-LAP antibody decreased tumor growth in mouse models of melanoma, colorectal carcinoma, and glioblastoma. Consistent with how the checkpoint inhibitors worked, which won James Allison a Nobel Prize, the anti-LAP antibody worked by enhancing antitumor immune responses. This time we published our results in *Science Immunology*, a very high-level journal, in a paper titled "Targeting Latency-Associated Peptide Promotes Anti-Tumor Immunity." Anti-LAP released the brakes on the immune system and allowed T cells to invade and destroy the tumor.

We studied two models of glioblastoma, one where the cells were injected in the flanks of the mice and another where the glioblastoma was injected directly into the brain. MRIs of the brains of these animals showed that anti-LAP treatment decreased tumor growth. In patients with glioblastoma, the expression of LAP and TGF-beta was associated with a poor prognosis. Our anti-LAP antibody was an attractive candidate for cancer immunotherapy.

We then took the anti-LAP experiments a step further by testing our monoclonal antibody with dendritic cell vaccination, which was being explored in patients with glioblastoma. Dendritic cells are powerful immune-stimulating cells. Scientists are using them to boost antitumor immune responses by putting tumor proteins on the surface of the dendritic cells and then immunizing mice with them. We injected glioblastoma cells that had the protein albumin on their surface into the brains of mice and then immunized mice with dendritic cells to stimulate T cells. We found a dramatic decrease of glioblastoma tumor growth in mice when we also treated them with our anti-LAP antibody. This showed that anti-LAP treatment could combine with dendritic cell vaccination.

Almost twenty years from the time we tried to induce TGF-beta-secreting cells to treat autoimmunity by oral tolerance and started a company called Autoimmune, Inc. that failed, we formed a new company called Tilos Therapeutics to commercialize anti-LAP antibodies that boosted the immune response to treat cancer. Barbara Fox became the CEO, and we transferred Taka's clones to Tilos for trials in humans. The next steps were now commercial: characterizing Taka's antibodies, turning the mouse antibodies into human antibodies, and performing clinical trials. Barbara Fox, who has a PhD in chemistry from MIT and has worked in higher management levels of several other pharmaceutical companies, told me she was excited about the science because more and more evidence was accumulating that TGF-beta was important in cancer and modulating it with anti-LAP antibody could make a big difference. She confessed to me that her first priority as CEO was to make sure she could repeat in an industry setting what I had done in the academic laboratory.

The goals of academic research and drug company research are fundamentally different. In academics, the goal is the discovery of new knowledge for knowledge's sake; in industry, the goal is turning knowledge into a commercial product. Both are needed to have new treatments for disease. Many times, what is done in an academic laboratory cannot consistently be repeated by others, even if the initial work is correct and accurate. Barbara said that it is not uncommon for the pharmaceutical industry to be unable to reproduce what happened in an academic laboratory, typically because the academic experiments had an unrecognized artifact or were performed under too stringent conditions. Fortunately, Tilos was able to exactly reproduce what we had done in my laboratory at the Ann Romney Center.

Once that hurdle was overcome, more money was raised, and the program was on its way. It takes hundreds of millions of dollars to bring a drug to market. Some people say as much as a billion. The funding stages for the launch of Tilos as a company were $2 million, followed by increasing amounts. Tilos continued to grow as the company hired more people and prepared for initial trials in people with cancer. I take pleasure knowing there are nine publications from my laboratory that serve as the backbone of Tilos's technology.

When I asked Barbara what she was most excited about by being the CEO of Tilos, she didn't hesitate in her response. "The chance to help people," she said. When I asked her what she hates the most, she said, "Having to raise the money and talking to investors. They always want things to go faster. They always want guarantees that everything will work." I asked her if she was afraid that anti-LAP antibodies might not work in glioblastoma or other cancers. She said she was not, because those were the rules of the game. In fact, most start-up companies that focus on new drugs for difficult-to-treat diseases ultimately fail. But to be able to have a shot on goal is all one can ask for. I agree with her; indeed, what we are trying to do at the Ann Romney Center is to come up with as many shots on goal as possible for the neurologic diseases we are fighting, knowing that one of them is bound to go in.

As it turned out, it became increasingly clear to the scientific community that targeting TGF-beta and LAP on the surface of cells was a major avenue for the treatment of cancer. More importantly, it became clear to the pharmaceutical industry. I remember the day I received a call from our hospital's innovation office that Merck would be acquiring Tilos for "a consideration up to $773 million," including an up-front payment and various milestones to develop anti-LAP antibodies for the treatment of cancer. A major shot on goal.

It is ironic that after spending so much time in my career on oral tolerance for autoimmune diseases, I came up with a treatment that might help treat cancer patients. When I discussed this with my wife, she reminded me of one of our favorite lines from a movie I wrote and directed, called *Abe & Phil's Last Poker Game*, starring Martin Landau and Paul Sorvino. In the film, an old Jewish doctor meets an old Italian womanizer and gambler in a nursing home. They strike up an improbable friendship and fight against old age. The head of the nursing home, who is a failed researcher, is trying to devise a potion that will cure old age. His idea is to feed people cancer cells with the idea that the cancer cells live forever and could lead to immortality. He tells Paul Sorvino about it, and Sorvino tells the nursing home director that he has cancer. The nursing home director thinks for a moment and then says, "Well, this treatment could not only stop you from aging, it could cure cancer on the side." Anyway, that's what happened to me after working for many years in MS: in the end, I found a potential treatment for cancer on the side.

At showings of my film, I am frequently asked whether feeding people cancer cells, as shown in the film, could really treat cancer. I tell them it would actually have the opposite effect and make cancer worse. Because of the phenomenon of oral tolerance, it would dampen the immune response to the cancer cells and the cancer would grow faster. Of course, few people who watch the film know enough about oral tolerance to realize this.

CURING GLIOBLASTOMA

Of all the brain diseases, glioblastoma is one of the most virulent and difficult to treat. This is largely because there are no major genetic influences or environmental processes impacting the disease that could be targeted. Accordingly, there is nothing to prevent it. Once it appears in the brain, it is late in the disease process.

Tracy Batchelor, who chairs the neurology department at the Brigham and Women's Hospital, has devoted his career to understanding and treating brain tumors. He has had the satisfaction of treating patients with lymphoma in the brain who can effectively be cured if treated early. One problem, however, is damage that may occur to the brain after therapy. This is a complication that will plague us as we one day find treatments that will cure glioblastoma. Tracy is studying IDH1 genetic mutations in glioblastoma, which make the tumors more resistant. He has identified candidate drugs that may specifically target cells that express IDH1 while sparing normal cells. He hopes we find an Achilles' heel in glioblastoma that can be targeted.

I asked Nino Chiocca, David Reardon, and Tracy Batchelor if people will still have glioblastomas a hundred years from now. They all believe there will be a cure, but finding it will be a slow, iterative process. The key is likely to be combination therapy that attacks different aspects of the tumor's growth, along with developing unique approaches to arm the immune system to destroy the cancer cells. Already, some long-tail survivors give us hope. With checkpoint inhibitors, oncolytic viral therapy, vaccination, and perhaps using what we know about the microbiome, more and more people will survive longer. Breakthroughs will come from better understanding unique properties of the glioblastoma cells and the cells surrounding them, including microglia and macrophages. It may be that powerful CAR T cells will be targeted to the areas of the tumor. I knew that Rob Bretholz would not survive when he came down with his glioblastoma, but I am certain that when I meet someone like Rob in the future, we will be able to cure them.

CHAPTER SEVEN

Going for the Cure

The word *cure* plagues medicine, and I must admit that it plagues me as well. I don't know how many times I've had a patient ask me when there will be a cure for the neurological diseases I've written about in this book. I have been fortunate to observe the development of drugs that have provided a cure on some level for some patients with relapsing forms of MS. If one of my family members came down with MS, I would feel confident that we have treatments that can keep the disease at bay so that they could live a normal life, raise their families, and grow old. I cannot yet say the same for Alzheimer's, ALS, or Parkinson's, and certainly not for glioblastoma. I have personally experienced what it's like to have a loved one or a dear friend with these diseases.

The word *cure* has almost become a cliché, especially in marketing and fund-raising. In the strictest sense, the word *cure* means to eradicate or to totally eliminate. It may be most accurately applied to a disease such as cancer or an infectious disease. If someone has a tumor and it can be removed by surgery, and there is no recurrence of the tumor, we can say that the person has been cured. If someone has a bacterial infection, it can be cured with antibiotics, depending, of course, on the bacteria. Polio was cured because of a vaccine to prevent it.

Cancer is definable because there is an ultimate endpoint to the disease, namely death. We even put numbers on it, such as a five- or ten-year cure rate. In general, we believe that if the cancer doesn't reappear in five years, one is cured. These concepts regarding cancer also apply to cancer of the brain, such as glioblastoma, but they don't necessarily apply when it comes to the other neurological diseases of the brain, which are generally progressive in nature.

Stopping the progression of ALS, Alzheimer's disease, or Parkinson's disease cannot truly be considered a cure, as the patient will be left with damage to the brain. Some of my MS patients, for whom we have successfully stopped the disease but who are left with disability, ask me, "When will there be a cure?" In these instances, they are referring to reversing the neurological damage they've acquired. They wonder about therapies to rebuild the brain so they can walk normally again.

Unfortunately, reversing neurological damage is extremely difficult because of the complexity of the nervous system and the difficulty of rebuilding the complex neural connections required for normal brain function. Maybe we should be careful about using the word *cure* when it comes to progressive neurological diseases and say "stop it in its tracks" instead.

The word *progressive* is our biggest enemy when we confront neurological disease. *Progressive* implies the inexorable movement of a biologic process over time. All of biology entails the progression of processes that have been set into motion: the birth of a baby after a sperm and egg have come together, a child's movement into puberty after reaching ten to thirteen years of age, and, of course, the movement toward infirmity and loss of biologic function as we age. Will we one day stop this natural biologic progression of aging? Is aging itself a disease and a process that we someday will cure?

MS is a unique disease because it consists of two processes: it has both relapsing-remitting and progressive stages. The other neurological diseases explored in this book are also inherently progressive in nature. In the beginning of the disease most MS patients have relapses followed by remission. If untreated, there is an attack that resolves, only to be followed by another attack that occurs months to years later. An MS patient who currently has symptoms such as numbness and coordination problems may have had an attack such as blindness ten or years earlier. When the first attack occurs, it is called a clinically isolated syndrome. I sometimes tell my patients it is "singular sclerosis." Nonetheless, when a person has a single clinical event, we often see on the MRI that much more is happening. One of the debates concerning MS is whether a person should be treated very aggressively when symptoms of the disease first appear, analogous to the treatment of cancer. Some conceptualize MS as similar to a cancer and believe that we should "cut it out" when it first appears with strong immune therapy. We know that if MS isn't treated or is not treated effectively, the majority of patients enter a progressive phase, even if they don't initially experience the progression happening. While it may take years, MS then becomes like the other progressive neurological diseases we are trying to cure—or, better stated, stop in their tracks.

Stopping a disease in its tracks means identifying the abnormal biologic process behind it and removing it. This is easily done if there is a single abnormality or a single gene that is defective. In the case of Parkinson's disease and Alzheimer's disease, one could argue it is the abnormal proteins that drive the diseases and stopping them in their tracks means preventing the accumulation of alpha-synuclein or A-beta. This is easier said than done, but at least we have specific targets, although as discussed, there is a debate among scientists as to what the targets really are.

The ultimate cure, of course, is to prevent the disease from happening in the first place. That's how we cured polio. In a sense, however, we didn't actually cure the disease itself, but we prevented it from ever happening. This may be a question of semantics: If you stop a disease from happening, did you cure it, or is it only cured if you already have the disease and then wipe it out? Let's refer again to the boxer analogy I give to my MS patients.

You have MS and you are a boxer in the ring with the disease. The cure would be to take you out of the ring, something we can't do yet. The ultimate cure is not to ever have been in the ring boxing with the MS boxer in the first place.

People who suffer disabilities from these neurological diseases react differently. Some say, "Can you find a cure?" Some say, "If you can just stop it, Doctor, I would be happy." Others say, "Can't you just slow it down?" Maybe the incremental steps are: slow it down; stop it in its tracks; cure it. A true cure may mean cutting it out completely or never letting it happen in the first place. Until we can rebuild the nervous system once the disease starts, maybe there is no cure. However, to say that once a disease starts there can be no cure does not fit with the fund-raising and marketing efforts needed to fight these devastating diseases.

People want hope. They want a big goal. People want the hero to slay the dragon and ride off in the sunset. Many often equate curing a disease with putting a man on the moon. Many patients have said to me, "If we could put a man on the moon, why can't we cure this disease?" Nixon had his war on cancer, and every five or ten years, there's a new war on a particular disease. The media reports breakthroughs as breaking news. Some of these breakthroughs really are magnificent: A child with spinal muscular atrophy who would be dead is now walking. Jimmy Carter would have died from his metastatic melanoma, but a new checkpoint inhibitor essentially cured him of his disease.

There is no question that we will eventually cure these diseases. But we still need to determine what the path will be, how we will get there, and what we can do to get there as quickly as possible. When patients bring me articles hyping new treatments for MS, ALS, or Alzheimer's based on research in a mouse or test tube, I am forced to tell them that it will take time to see if the treatment will work in humans. It's frustrating both for me and for those with the disease, but it can take decades for new discoveries or individual treatments to succeed in the clinic. One study found that from the time a discovery is made, it takes a minimum of twenty years before it becomes a treatment for people.

In the media and in marketing hype we usually look only five or ten years ahead. I fall prey to this myself. It's hard to tell a patient, "Well, we will have a cure in twenty-five years," although we know that it usually takes that long for all the advances to be made. The National Institutes of Health usually give grants for only five years, and programs for therapy are spoken of in those terms. When we say we will have a cure in five or ten years, maybe we can't be honest because it's so hard to take away a patient's hope.

I've taken care of individual MS patients for decades, and they sometimes confront me with this: "Hey, Doc, five years ago you said we'd have a cure or new treatment in just five years' time, but the five years keep going by." In the past two decades, we have had new treatments in five-year blocks of time, but no cure and nothing to rebuild the nervous system.

Curing any disease requires understanding the biologic processes that drive it and finding ways to stop them. It is not any more complicated than that. What *is* complicated are the biologic processes themselves. It often is not a single event or process that drives disease but a combination of them. It is like the analogy of an avalanche. Once it has started, it's hard to stop. Can we understand what starts the avalanche? If so, can we stop it from happening? We don't want to be trying to cure disease once a mountain of snow is hurtling down the mountain. Again, when it comes to confronting an individual patient, it is so difficult for us to say, "Nothing more can be done."

Basic research that leads to an understanding of the fundamental processes by which biology works and how it goes astray in disease is the ultimate key to curing diseases. This is not sexy, and it's not something that the disease-focused societies and organizations can raise money for. It's hard to tell someone who wants their loved one with Alzheimer's disease to have a therapy as soon as possible that we need to invest money in zebrafish or studies of proteins and how they fold or don't fold in a dish in a laboratory.

In truth, curing a disease requires support on many, many levels. Translational research that can be extrapolated from lab or animal studies is clearly needed, but it bears repeating that ultimate cures will come from basic science and understanding the basic processes. Support of basic science is now done through the NIH and other research institutes. One could argue that disease-focused societies such as the National MS Society and the Michael J. Fox Foundation for Parkinson's Research should set aside a certain percentage of the funds they raise for basic research that is directly relevant to the disease of interest, such as the study of essential mechanisms of brain function. I can tell you, however, this is a difficult sell to the donors who give money for cures. They want direct attacks on the diseases. What is needed, then, is the ability to understand and translate basic scientific advances of any kind into comprehension of a specific disease and new therapies to treat it. The NIH has recently used this approach in Alzheimer's disease by offering supplemental funding to anyone who has an NIH grant independent of Alzheimer's and can apply what they are studying to the understanding of Alzheimer's disease.

There are many centers trying to understand and treat neurological diseases. At the Ann Romney Center, Dennis Selkoe and I have tried to create a plan of attack on these neurological diseases that is multifaceted and factors in all the steps needed for success. At the core of our efforts are a critical mass of the very best scientists who feed off each other as they meet in the hallway or lunchroom and come up with ideas and collaborations. The foundation of our center rests on scientists pursuing basic questions related to the brain, funded by very competitive NIH grants. Each year, I send Ann Romney a video from our holiday party with over three hundred scientists eating and actively talking science. They stop to wave at her and then return to their discussions. It is visual evidence of the mass of energy working on these diseases.

As I mentioned at the opening of this book, when Ann and I first conceived of the Ann Romney Center, she was in the middle of her husband Mitt's presidential campaign and said to me, "Howard, you should think about specific slogans or touchstones that can explain and drive what we're doing." I came up with the three I cited in chapter one: Drilling for Oil, Breaking Down Silos, and Shots on Goal. Over the years, I've grown to like them more and more. They have served as guideposts for the steps we must take to first slow a disease, then stop it, and ultimately cure it. These concepts apply to all of scientific discovery.

The first, Drilling for Oil, encourages our scientists—and this would apply to scientists everywhere—to drill for oil or look for new areas of discovery knowing that most potential oil wells are dry. Once scientists have the basic funding to investigate well-reasoned questions, they have to be encouraged to drill for oil. This is where philanthropic and other funds are important.

The second guidepost, Breaking Down the Silos, counters one of the natural consequences of science and how it's performed, which is that people work in their own specialized areas. Often, when two disparate technologies or people working in separate fields are brought together, ideas or concepts that weren't thought of before emerge. We only know what we know. The question is, How do we learn about the things we don't know? Breaking down silos occurs not only in applying new technologies to an old problem but also in collaborating with scientists from disparate labs and disciplines. We endeavor to do this within our center, as well as across different hospitals and institutions in the United States and around the world.

The third and final guidepost is Shots on Goal. I believe we always must be testing new concepts of therapy in patients who have the disease. Ultimately, we want to sit across from a patient and say, "We have a treatment we're going to try. We have a new therapy we are advancing." You can't score a goal if you don't take a shot on goal. We know that most shots fail, but every shot not taken is a failed shot. Animal studies are needed to understand basic mechanisms and to test treatments for both safety and efficacy before they can be given to people. Without animal studies, we wouldn't have most of the treatments that are currently in use. Nonetheless, a true shot on goal requires studying and treating those with the illness and then learning from what is discovered. The animals we treat are genetically identical, but people are much more complex, and moving from the laboratory and animals to people is very difficult.

We need to be taking more shots on goal, and we do this at the Ann Romney Center. An important aspect of the shots on goal concept relates to serendipity and the unexpected. Many scientific discoveries have been made by chance, where something unexpected led to a major discovery. When we take a shot on goal and test a new therapy, we may learn something unexpected about the disease.

The pressure on the investigators at our center is intense and relates to both time and discovery. To survive academically, we must write grant applications and publish papers, preferably in high-level journals, because grants, which usually take more than one nine-month cycle to be funded, require data that have already been gathered—an oil well that is already producing oil. The publication of findings in a top-level journal often takes two to three years of complicated experiments and then a year of additional experiments to satisfy reviewers. We have to deal with cost cutting, whether caused by the NIH making across-the-board cuts, a sudden rise in the cost of research animals, or the additional expense of performing sophisticated gene expression analysis. Philanthropic or institutional funds buy time to allow discovery to happen. As academics, we need patrons, just as artists do. We, however, are not trying to create beautiful pictures or music, we're trying to help people live more beautiful lives.

In order to take a successful shot on goal, much more than research is required to bring a treatment to a wide number of patients. There are the complexities of forming biotech companies and enticing the pharmaceutical industry to invest millions of dollars in a potential new drug. We have seen how much it costs to bring a drug to market, the costly trials that need to be performed, how many trials have failed, and the navigation that is required at the FDA. The culture at drug companies and for investors often goes counter to the freedom of time that is needed for breakthroughs. The drug industry makes reference to "The Valley of Death," meaning the valley that must be crossed in order to test a drug in people. We can help bridge the valley by our shots on goal, and we always need to patent our discoveries so that investors have protection for their investment. The challenge at all levels relates to having sufficient funds, whether it is for basic research, starting a biotech company, or doing a multimillion-dollar Phase 3 trial.

Dennis and I purposely named our center a Center for Neurologic Diseases, not a Neuroscience Institute. We specifically chose the word *diseases* because we are sensitive to patients and our desire to treat them. Our center is located in the Hale Building for Transformative Medicine at the Brigham and Women's Hospital, a magnificent building with imaging and patient examination rooms on the lower floors, and clinically related lab research and basic lab research on the upper floors. Experimental animals are on the very top floor, contrary to usual practice; animals are usually placed in the basements of buildings where research is performed. We needed the ground floor for the twenty-five-ton MRI magnets, including a new 7 Tesla magnet, the first clinical 7T magnet in the country. Our building is a center for translational research. We are always thinking of ways in which the basic research that's being done can be translated into a clinical study to treat the disease. As I said, we of course are not the only center working on these diseases, but I do believe we are unique in many ways.

THE IMMUNE SYSTEM AND THE BRAIN

As we have seen, the immune system plays a major role in the propagation of disease in the brain, as well as in its treatment. In many ways the immune system and the brain are alike; they evolved to deal with a complex environment, and they both have memory. The immune system is also an integral part of the crime scenes, not only of the five diseases I have discussed but with other diseases affecting the brain, too. Perhaps I'm viewing brain disease through the lens of my chosen discipline of immunology, but indeed the immune system appears to be central to understanding the treatment of brain diseases.

Three major components of the immune system relate to brain diseases. The first component is T cells, which are at the heart of what causes MS. The movement of T cells into the brain initiates the process. T cells have also been reported to react with A-beta in Alzheimer's disease and with alpha-synuclein in Parkinson's disease. In ALS, there appears to be a defect of regulatory T cells, so treatments, including some of ours, are being designed to enhance regulatory T cells in ALS patients. In brain tumors, there is a shortage of killer T cells to fight the tumor, and specially engineered CAR T cells are being used to treat cancer and are being considered as treatment for autoimmune and other diseases.

The second major component of the immune system that relates to brain diseases is antibodies. Antibodies directed at abnormal brain proteins can be administered systemically and are able to enter the brain and clear the abnormal proteins. This approach is being pursued to treat Alzheimer's disease using anti-A-beta antibodies, and a similar antibody approach is ongoing in Parkinson's disease with testing of antibodies directed against alpha-synuclein.

The third major component of the immune system that relates to brain diseases is the specialized immune system in the brain, made up of microglial cells and astrocytes. There has been a revolution in our understanding of microglial cells, and it appears that they play a major role in almost all diseases in the brain: they become inflamed, they become toxic, and they lose their protective function. The same can be said for astrocytes, which can be both good and bad. Targeting microglial cells and astrocytes could have a major impact on brain disease.

I have focused on five diseases in this book, but there are other diseases of the brain that may also involve the immune system. One is epilepsy. It can be effectively treated in the majority of cases, but other people's seizures are not well controlled. The most common cause of epilepsy is a traumatic insult that leaves a damaged area in the brain, which serves as an epileptic focus. This initial insult may leave localized inflammation that could be a major factor in the difficulty in controlling the seizures. This is also true for stroke; the secondary inflammatory immune response makes stroke worse. The immune approaches we are developing could ultimately be applied to both epilepsy and stroke. Finally, in traumatic brain injury, the head trauma initiates a cascade of events that involves activated microglial cells, and this, too, may be treated by dampening the local immune response in the brain.

PATHWAYS OF INVESTIGATION

We may not know exactly where new basic discoveries will come from, but certain areas that will clearly impact our ability to slow down, stop, and cure disease have opened up. One such area is the microbiome. As we have seen, the gut microbiome appears to play an important role in all the diseases we have discussed: MS, ALS, Alzheimer's disease, Parkinson's disease, and glioblastoma. Epilepsy has also now been linked to the gut, with data showing that a ketogenic diet may have benefits in that condition due to substances in the gut that then affect the brain. The gut-brain connection is a powerful one that is only now beginning to be understood. We have seen that it is possible that Parkinson's disease starts in the gut, with constipation being the first symptom.

The gut affects the brain through a number of pathways. First, neurotransmitters produced by intestinal cells can signal through nerve fibers in the vagus nerve or spinal cord, the nerves that in turn affect the brain. Not only can intestinal cells produce neurotransmitters but bacteria themselves can also produce neurotransmitters. These neurotransmitters can also travel directly to the brain from the gut. There is signalling that goes from the gut to the brain, and from the brain to the gut. Metabolic mediators of brain function, such as short-chain fatty acids, may affect microglial cells in the brain. Tryptophan metabolites in the gut could be linked to depression and cognitive function. Second, bacteria in the gut can modulate immune cells, which can migrate to the brain. And third, bacteria in the gut may have structures that are similar to structures in the brain. When there is an immune response against the bacteria, this triggers an attack on the brain structures that are shared between the bacteria and the brain.

I cannot imagine that our ability to control, stop, and cure neurological diseases will not somehow involve the gut. I imagine that in the future, both blood and stool samples will be analyzed for possible indicators of brain disease, and that we will devise more and more sophisticated ways to manipulate the gut. The field is in its infancy, but understanding the mechanisms by which the gut affects the brain will ultimately lead to treatments.

Other areas of advancement will involve more sophisticated methods of imaging the brain so that we can follow the progression of disease in living individuals. Linked to this are identifying biomarkers and developing precision medicine, establishing ways to understand which treatment a person should receive and those who are at risk for disease. Getting to this point will involve a complex integration of thousands of data points from blood, urine, stool, brain, and saliva samples, which are then analyzed, perhaps using artificial intelligence.

Industry is actively trying to develop sophisticated programs for personalized medicine. In MS, we are collaborating with industrial partners, who take advantage of the thousands of our CLIMB longitudinal cohort study samples. At the Ann Romney Center, the neurogenomics laboratory of Clemens Scherzer is accumulating a large databank of samples from patients with Parkinson's disease, and Dennis Selkoe is developing blood tests for Alzheimer's disease, as it isn't practical to perform PET imaging for amyloid on everybody. Wearable sensors will help track the daily activities of patients and the physical manifestations of brain function.

TELLING THE STORY OF THE SEARCH FOR A CURE

I always tell my fellows that the best scientific talks, the best grant proposals, and the best scientific experiments all explore a narrative. I have been very interested in how the story of disease is told. In the two movies I have made, *What is Life? The Movie*, which deals with life's big questions, and

Abe and Phil's Last Poker Game, which deals with end-of-life issues, I've turned to people with disease to understand how their stories are told. In *What is Life? The Movie*, one of the most poignant moments is when I ask a couple what they fear most. The husband, who is caring for his wife with MS, says that he fears most that she will die sooner than he does. In *Abe and Phil's Last Poker Game*, the old doctor who enters the nursing home is there because he can't take care of his ailing wife, who has Alzheimer's. The actress who played Abe's wife, Molly, had few speaking lines and like my mother conveyed the loss of connection and understanding of the world around her by a blank look on her face. The actress jokingly complained that here she was acting opposite the great Martin Landau but had few speaking lines.

Hollywood recognizes the power of narrative as it relates to disease and has made compelling feature films about people facing Alzheimer's, ALS, and Parkinson's. Julianne Moore received an Academy Award for portraying a woman with genetically dominant Alzheimer's disease in *Still Alice.* Eddie Redmayne received an Academy Award for his portrayal of Stephen Hawking and his fight with ALS in the movie *The Theory of Everything.* Academy Award winner Hilary Swank played a pianist with ALS in *You're Not You*, while Robert De Niro played a patient with Parkinson's disease in *The Awakening.* Emily Watson was nominated for an Academy Award for her portrayal of Jacqueline du Pré, a cellist with MS, in *Hilary and Jackie.*

The narrative of disease is usually told in one of two ways. The first narrative depicts the courage those with the disease and their families must have, and their yearning for something to be done. The second narrative highlights disease as a success story in which the puzzle is solved, and the disease is cured. Documentaries on the search for a cure focus on people undergoing therapy in real time. These stories are poignant, but too many of them do not end happily.

I've been especially struck by the story of Simon Fitzmaurice, who succumbed to ALS after writing about his struggle in the book *It's Not Yet Dark.* As discussed, he was a remarkable man with a wife and five children.

He fulfilled his dream of writing and then directing a movie, although he directed the movie from a wheelchair while completely paralyzed, giving directions with his eye movements. I cannot honestly say that I would have that much courage. But I would probably fight by trying to come up with the treatment for myself despite the odds. Ralph Steinman, who won the Nobel Prize in Medicine for discovery of dendritic cells in the immune system, did just that when he developed pancreatic cancer and devised a treatment for himself based on the discoveries he had made—a dendritic cell vaccine. The treatment may indeed have briefly prolonged his life before he finally succumbed to the disease. Unfortunately, he died three days before his Nobel Prize was announced. Although Nobel Prizes are never given to those who are not alive, the committee had not realized that he had passed, and his Nobel Prize was awarded.

How important are prizes such as the Nobel Prize for scientific advancement and ultimately curing disease? From what I've observed, those working to solve these diseases are primarily driven by their curiosity and their interest in the science, not in being awarded a particular prize. Everyone is touched by the plight of people with untreatable diseases and the opportunity to help them. It is possible for someone to decide that they will make a fortune or, if they are talented, become a star athlete. However, it is almost impossible to set out to win a Nobel Prize; so many of the prizes come from serendipity and the unexpected. Nonetheless, it's always in the back of everyone's minds.

Will it help scientific progress to make scientists and doctors heroes? It actually might have the opposite effect by interfering with the science, though if it encourages young people to enter the field, it would be a good thing. What will most help scientists and doctors is to make sure they have enough money to do their work.

The brain is unique, and because of this it poses unique challenges. Its ultimate uniqueness is being the seat of the soul, the organ that makes us who we are. I often kid my cardiology colleagues that the heart exists simply to pump blood to the brain. When I went to my orthopedic doctor because of pain in my knee, he said, "Howard, it's just your knee, we can replace it. Remember, it's not your brain." Some think that far into the future we'll be able to download our consciousness into a computer and live forever, and in effect to have an artificial brain as we now have an artificial knee or kidney. Then, there would be no brain diseases to cure.

Of all the diseases I've written about in this book, I must admit that Alzheimer's is our number-one challenge and the number-one scourge of civilization at this time. It robs us of who we are, and it increases exponentially with age. Society can't afford not to find the treatment for Alzheimer's disease within the next decades. The cost of care will be staggering; we will be unable to care for all those with the disease.

The blueprint for slowing, stopping, and then curing Alzheimer's will be early detection and early therapy and then ultimately therapy to prevent the disease from happening. With the initial positive results of the anti-A-beta antibody trials, we hope this will ultimately become a major milestone that will energize the field and put us on the road to effective therapy. There is no choice but to find treatments for Alzheimer's disease that will work.

How will we cure the five brain diseases I discuss in this book? The cure for Alzheimer's disease will involve a series of steps taken in middle-aged people to protect the brain. It will involve not only antibodies and cellular vaccines but also other approaches, such as manipulating the microbiome and genetic identification of those at higher risks as part of personalized medicine. We will cure MS by treating the disease early to stop the process and by developing ways to prevent it. ALS will eventually succumb to approaches that correct gene abnormalities, protect neurons, and shut down the inflammation that amplifies the disease. The cure for Parkinson's disease will rely on early detection and initiating treatment before motor systems are involved, as well as targeting alpha-synuclein. Glioblastoma may be the most difficult to cure because there's no way to prevent the tumor. A cure will most likely involve immune therapy.

THE FUTURE

What is it like to be in the trenches at the Ann Romney Center? It is both exhilarating and frustrating. It is a struggle to perform experiments with funding restrictions or without funds to pursue a crazy idea. The exhilaration comes from interacting with colleagues, seeing improvement in patients with a disease, and designing the next experiment. I always finish my conversations with scientists at our center with the words "Keep going, keep going." I try to raise money so that when someone needs extra funds or loses a grant, I never have to say no and can offer help. I realize how great the odds are against us, but I don't confide that readily to the young scientists working on the diseases.

What should you do if you contract one of these diseases? Depending on the disease and the stage of the illness, there may be a treatment, but otherwise, it's about marshalling the courage to confront the illness and appreciate life as much as possible. Many of my patients feel it is somehow their fault for coming down with the disease. However, no one is at fault. It is simple bad luck. Individuals can contribute by participating as subjects in research studies and clinical trials. My childhood friend Normie recently passed away from his MS. It was too late to help him. One day, the research will be successful, and we will have cures. Will we ever have a world free of disease? Not likely. Will we ever live forever? That is a bigger philosophical question.

I was born on Christmas Day in 1944. From the time I can remember, my mother told me the story of her father, who perished in the Holocaust. He always wanted a doctor in the family, and he wrote to her before he died to make sure her first child would be a doctor. When I was born, she told me that I was to be that doctor. I of course do not remember that moment, but I do remember hearing the story as a little boy. As I grew up, I realized I really did want to be a doctor. When I walked past a hospital, I somehow felt that I belonged there. In the end, I don't care whether I became a doctor because my grandfather told my mother to tell me to become a doctor or whether it was something that was part of me when I was born. Of course, I do have some of my grandfather's genes.

When I look back on my career so far, my greatest satisfaction has come from having helped people. I follow patients who we treated with chemotherapy for aggressive MS, and who decades later are stable and living normal lives. Luckily, I now have drugs to give to my MS patients. Sadly, there are no drugs for Alzheimer's. When my mother said to me, "Howard, you're a brain doctor, can't you cure my disease?" there was nothing I could do for her Alzheimer's and nothing I could do for Tanya's ALS. I now have hopes for the nasal vaccine, although I know the odds are against me. Nonetheless, there are other shots on goal I am preparing to take.

My biggest regret is that I don't have fifty more years to investigate these diseases, understand them better, and devise new treatments, and even more, to see that they have been cured. We all want more time, but our time is limited, and we have to be thankful for what we have. I am thankful that I have shots on goal coming up—the nasal vaccine for Alzheimer's that I told Ann Romney about, which triggered the formation of the Ann Romney Center. Oral and nasal anti-CD3 not only for progressive MS but for other diseases as well. Anti-LAP antibody for cancer. MicroRNAs for ALS. I am not naïve. I know how hard it is to succeed at a shot on goal in medicine. Nonetheless, I remain optimistic. It is said that you are not expected to always succeed at a daunting task, you only are expected to give your best effort. I don't consider myself a unique individual or special researcher. I consider myself part of the collective human curiosity and drive to make our world a better place. *Tikkun olam* in Hebrew means "repair the world," and I am happy to be a part of it.

I hope that you have been captivated by the amazing biology of these diseases, their complexity, and the long and often circuitous path to finding a cure. One doesn't have to travel to outer space to explore the uncharted waters of the biology of disease. I hope I have provided you with a firsthand account of the fight to find cures and how we are truly making progress. To those suffering from these diseases and to their family members, be assured there will be a cure someday—unfortunately, often not in the five- or ten-year time frames that are promised, and unfortunately, too late for many.

This is difficult for me as a doctor, as patients understandably want something done immediately. Few are comforted by the fact that we will have a treatment or a cure in ten or twenty years, although in the history of medicine a decade or two is not a long time. As for all biologic processes, there is a rhythm to discovery, and things take as long as they must take. I sometimes remind people that two women can't give birth to a baby in four-and-a-half months.

We are all on the same ship of humanity, sailing together. Take wonder in the biology of life, stay close to family and friends, and keep on going. My wife tells me that I am an unrivaled enthusiast, which is what I like to be. In my office, I have a number of signs and sayings about the scientific process—the importance of receiving honest criticism from colleagues, the need to persevere, to never quit, and to accept that life is not perfect and we must learn to deal with it. One that everybody gravitates to is a simple one that says, "Cure as many diseases as possible." This is what we're all trying to do.

GLOSSARY

Acetylcholine

An excitatory neurotransmitter or chemical messenger released by nerve cells to send signals to other cells. It is decreased in Alzheimer's disease. The drug Aricept is used to treat Alzheimer's disease by increasing levels of acetylcholine.

ADAS-Cog

Alzheimer's Disease Assessment Scale–Cognitive Subscale. A scale used to measure cognition in Alzheimer's disease.

Adjuvant

A compound that boosts the immune response.

Akkermansia

A bacteria found in the gut that is associated with disease, including multiple sclerosis and ALS.

Alpha-synuclein

A toxic protein that accumulates in brain neurons of patients with Parkinson's disease and makes up Lewy bodies.

ALSFRS-R

ALS Functional Rating Scale-Revised. A scale used for monitoring the progression of disability in patients with ALS.

Amino acids

Building blocks that combine to form proteins. There are twenty amino acids.

Amyloid-beta

A toxic protein that accumulates and forms plaques in the brains of patients with Alzheimer's disease. The amyloid hypothesis states that accumulation of amyloid is the cause of Alzheimer's disease, and treatments to remove amyloid are being tested.

Antibody

A Y-shaped structure that is part of the immune system. Antibodies bind specifically to their target and help fight infection. For instance, a polio vaccine induces antibodies that bind to and inactivate poliovirus but not a smallpox virus.

APOE4

The most prevalent genetic risk factor for Alzheimer's disease.

APP gene

A gene that encodes the amyloid-beta precursor protein (APP). Mutations in the *APP* gene lead to familial forms of Alzheimer's disease. Cleaving the *APP* protein leads to the accumulation of toxic amyloid-beta protein in Alzheimer's disease.

ASO

Antisense oligonucleotides (ASO) are small, single-stranded molecules of DNA. They are being used therapeutically because they can enter a cell and dial down an abnormal gene.

Astrocyte

A major cell type in the brain that provides support for the brain both structurally and metabolically. It can be both protective or toxic and can become a brain tumor when it grows out of control.

Autoimmune disease

A disease in which the body's immune system goes awry and attacks itself. Examples are multiple sclerosis, type 1 diabetes, and rheumatoid arthritis.

Autophagy

A mechanism by which a cell removes unnecessary or dysfunctional cell components. It helps protect the cell but can also promote cell death and morbidity.

Axon

Nerve fiber in the brain and spinal cord that carries electricity.

B cell

A white blood cell that is part of the immune system. It is the immune cell that makes antibodies. Monoclonal antibodies can be made from immortalized B cells. Targeting B cells is a treatment for multiple sclerosis.

Beta-amyloid

See amyloid-beta.

Biomarker

A marker of a biologic process that is used in research to understand, monitor, and screen for disease. Used in blood tests, imaging studies, and genetics.

Blood-brain barrier

A border that prevents entry of substances from the blood into the brain.

C9orf72

A major genetic risk factor for ALS.

Checkpoint inhibitors

Drugs used to treat cancer by targeting the immune system and releasing checkpoints, which then allows the immune system to attack cancer cells.

Chelation

A process by which toxic heavy metals are removed from the body.

CLIMB study

Comprehensive Longitudinal Investigation of Multiple Sclerosis at the Brigham and Women's Hospital.

CNS

Central nervous system. The brain and spinal cord, as opposed to the peripheral nervous system, which consists of nerves outside the CNS that connect the CNS to the rest of the body.

CSF

Cerebrospinal fluid. The fluid bathing the brain and spinal cord.

Cyclophosphamide

A chemotherapy drug used to treat multiple sclerosis.

Dendritic cells

Accessory immune cells, which trigger immune responses by T cells.

Dopamine

A neurotransmitter or chemical messenger released by a nerve cell. A deficiency of dopamine is the major cause of symptoms of Parkinson's disease.

Double-blind trial

A clinical trial in which neither the patient nor the doctor knows which treatment the patient is receiving.

EAE

Experimental allergic encephalomyelitis is a widely used animal model for multiple sclerosis.

EDSS

Expanded disability status scale. A clinical scale to measure disability in multiple sclerosis.

Enzyme

A protein that accelerates chemical reactions in biological systems.

Epstein-Barr virus

A virus in the Herpesviridae family that is the cause of infectious mononucleosis.

Fecal microbiome transplant (FMT)

A stool transplant in which the stool of a healthy individual is transferred into another individual.

GABA

Gamma aminobutyric acid. A neurotransmitter or chemical messenger that blocks impulses between nerve cells in the brain.

Gadolinium

A substance injected into the vein at the time of an MRI. If there is brain inflammation, gadolinium leaks into the brain and lights up on MRI (gadolinium enhancement).

Glutamate

A neurotransmitter or chemical messenger that is a major excitatory neurotransmitter in the brain.

Gut-brain axis

The bidirectional communication between the gut and the central nervous system. Bacteria in the gut (microbiome) may affect the gut-brain axis.

Idiopathic

A disease for which the cause is unknown.

Interferon

A natural substance that interferes with the replication of viruses and also affects the immune system. Interferon drugs are used to treat multiple sclerosis.

IVIG

Intravenous immunoglobulin. Antibodies purified from human blood that are used to treat immune disorders.

LAP/anti-LAP

Latency associated peptide. A protein on the surface of an immune cell that contains the suppressive chemical TGF-beta. Anti-LAP is an antibody directed against LAP that is a checkpoint inhibitor to treat cancer.

L-DOPA

The precursor to the neurotransmitter dopamine, used to treat Parkinson's disease.

Lewy bodies

Abnormal aggregates of the protein alpha-synuclein that accumulate in the brain of people with Parkinson's disease.

LRRK2

A major genetic risk factor for Parkinson's disease.

Lymphocyte

A type of white blood cell that is part of the body's immune system.

Macrophage

A white blood cell that has the property of engulfing particles; a scavenger cell.

Mast cell

A white blood cell that is part of the immune system. Mast cells produce histamine and are important in allergy.

MBP

Myelin basic protein. A protein of the myelin sheath that is damaged in multiple sclerosis.

Methylation

A biologic process by which methyl groups are added to DNA, which can change DNA function.

Microbiome

All the microbes (bacteria, fungi, protozoa, and viruses) that live in the human body. The gut microbiome refers to the organisms in the gut.

Microglial cell

Immune cell of the brain analogous to macrophage in other parts of the body. It helps maintain brain function and responds to injury and infection. It can be both protective or toxic.

Mitochondria

A component of the cell that provides energy for the cell.

Molecular mimicry

Structures found on viruses or bacteria that are identical to structures found on human cells. If the immune system attacks such a structure on infectious agents, it may then mistakenly attack a normal human cell that has the same structure.

Monoclonal antibody

An antibody made from B cells that have been immortalized so they produce unlimited amounts of a unique specific antibody in culture vessels. Monoclonal antibodies are used as drugs and in medical research.

Monocyte

A white blood cell in the bloodstream that becomes a macrophage when it leaves the blood; a scavenger cell that engulfs particles.

Motor neuron

A neuron located in the central nervous system whose axon (fiber) controls motor function throughout the body. Motor neurons are damaged in ALS.

MPTP

A neurotoxin that destroys dopamine neurons in the substantia nigra and causes symptoms of Parkinson's disease.

Myelin sheath

The insulating sheath that surrounds nerve fibers (axons) in the nervous system. Myelin is made by oligodendrocytes in the central nervous system. It is damaged in multiple sclerosis.

NEDA

No evidence of disease activity. A measure of the effectiveness of treatment in multiple sclerosis.

Neurofibrillary tangles

Aggregates of tau protein that are a major microscopic marker of Alzheimer's disease.

Neurofilament

A protein inside neurons that helps form the scaffolding of the cell. When neurons are damaged, it is released into the bloodstream and may serve as a measure of brain damage.

Neurons

The fundamental units of the brain and spinal cord that transmit information in the brain. They control motor function and receive sensory input. They communicate by electrical impulses and chemical messengers (neurotransmitters).

Neurotransmitter

A chemical messenger that is released at the end of a nerve fiber.

Oligoclonal band

Patterns of elevated gamma globulin in the spinal fluid that are seen in multiple sclerosis and used as a diagnostic test.

Oligodendrocyte

A cell in the central nervous system that makes myelin and that is damaged in multiple sclerosis.

Oncolytic virus therapy

A treatment used for brain cancer (glioblastoma) in which viruses are used to help stimulate the immune system to attack the cancer.

Optic neuritis

Inflammation of the optic nerve causing loss of vision. A common symptom of multiple sclerosis.

Oxidative stress

Dysfunction in a cell caused by the accumulation of toxic oxygen reactive species.

P value

A statistical test to determine the degree to which an outcome could have happened by chance. In biology, a P value of less than 0.05 is often considered significant, a one in twenty possibility that the result happened by chance.

Peripheral nervous system

The part of the nervous system that is outside the central nervous system (CNS) and connects the CNS to the rest of the body.

PET imaging

Positron emission tomography. An imaging technique that uses radioactive substances to visualize structures in the body, including microglial cells and amyloid in the brain.

Placebo

A sham pill or treatment.

PML

Progressive multifocal leukoencephalopathy. A rare and often fatal viral infection of the brain caused by the JC virus. PML occurs in immunosuppressed multiple sclerosis patients, especially in those treated with Tysabri.

Presenilin gene

A gene that provides instructions for making a protein called presenilin. Mutations in the presenilin gene are the most common cause of familial Alzheimer's disease.

Primary endpoint

In a clinical trial, the primary endpoint is the primary question being addressed by treatment. For a trial to succeed, the primary endpoint must be met.

Prion

An abnormal protein that can trigger normal proteins in the brain to fold abnormally and cause disease. Prions are proteins that can be infectious.

Prodromal

An early sign or symptom that indicates the onset of disease before a formal diagnosis is made.

Progressive MS

A form of multiple sclerosis that involves continued worsening of disease without intervening attacks.

Prostaglandins

Chemicals derived from lipids that have multiple physiologic functions in the body and can be increased during inflammation.

Relapsing remitting MS

The most common form of multiple sclerosis, which involves attacks of disease followed by recovery.

REM sleep

Rapid eye movement sleep. A stage of sleep in which dreams occur. Abnormal in Parkinson's disease.

RNA

Ribonucleic acid. It serves as a messenger, carrying instructions from DNA to control synthesis of proteins.

Schwann cell

A cell in the peripheral nervous system that makes myelin and surrounds nerve fibers.

SOD

Superoxide dismutase. An enzyme in the cell that breaks down toxic oxygen molecules. It is abnormal in ALS.

Stem cells

A primitive cell that can differentiate into many different cell types. Embryonic stem cells come from fetal tissue. Mesenchymal stem cells come from bone marrow. Induced pluripotential stem cells are made from skin or blood cells.

Substantia nigra

A dark pigmented structure in the brain that contains dopamine neurons and is abnormal in Parkinson's disease.

T cell

A white blood cell of the immune system that is the major effector cell and regulatory cell of the immune system. It is named for originating in the thymus. There are many types of T cells, such as Th1, Th2, Th17 cells, and Treg cells. T cells that attack the myelin sheath initiate multiple sclerosis.

Tau protein

A protein that helps stabilize neurons. In Alzheimer's disease the tau protein forms tangles in the neuron.

TGF-beta

Transforming growth factor beta is a chemical produced by white blood cells that suppresses immune responses.

Transgenic

An animal in which a foreign gene (a transgene) has been inserted.

Tumor necrosis factor (TNF)

A chemical released by inflammatory cells of the immune system.

Vagus nerve

A nerve that runs from the brain stem to the thorax and abdomen and is part of the gut-brain axis.

Ventricles

Cavities in the brain where spinal fluid is made.

RESOURCES: RESEARCH STUDIES, CLINICAL TRIALS, AND ORGANIZATIONS

Accelerating Medicines Partnership—Parkinson's Disease
https://amp-pd.org/

Aligning Science Across Parkinson's
https://parkinsonsroadmap.org/

ALS Association
https://www.als.org/

Alzforum
https://www.alzforum.org/

Alzheimer's Association
https://www.alz.org/

American Brain Tumor Association
https://www.abta.org/

Ann Romney Center for Neurologic Diseases at Brigham and Women's Hospital

https://www.brighamandwomens.org/research/departments/center-for-neurologic-diseases/ann-romney-center-for-neurologic-diseases

Clinical Trials

https://clinicaltrials.gov/

Federal Drug Administration (FDA)

https://www.fda.gov/home

Michael J. Fox Foundation for Parkinson's Research

https://www.michaeljfox.org/

National Institute of Neurological Disorders and Stroke

https://www.ninds.nih.gov/

National Institutes of Health (NIH)

https://www.nih.gov/

National Multiple Sclerosis Society

https://www.nationalmssociety.org/

Northeast ALS Consortium

https://www.neals.org/

Parkinson's Foundation

https://www.parkinson.org/

PatientsLikeMe

https://www.patientslikeme.com/

UsAgainstAlzheimer's

https://www.usagainstalzheimers.org/

ACKNOWLEDGMENTS

A large number of physicians and scientists have made major contributions to our understanding of brain disease. The people I refer to in this book are those I have worked with or have come in contact with and whose research related to the aspects of research chronicled in this book. There are countless others who are not mentioned who have played critical parts in advancing our understanding of brain disease.

I would like to thank those who gave of their time to be interviewed: Tracy Batchelor, James Berry, Robert Brown, Jr., Nino Chiocca, Merit Cudkowicz, Stephen Hauser, David Reardon, Al Sandrock, Clemens Scherzer, and Dennis Selkoe.

Those who provided critical comments on the manuscript: Carl Axelrod, Rohit Bakshi, Oleg Butovsky, Tanuja Chitnis, Laurie Cox, Leslie Epstein, David Fisher, Joe and Rhoda Kaplan, Vik Khurana, Vijay Kuchroo, Walter Klenhard, Evy and Joe Megerman, Francisco Quintana, and Dennis Selkoe.

Ellis Trevor and Michael Mungiello, my literary agents. David Martin, who helped with the early stages of the book. Alexa Stevenson and all those at BenBella Books who believed in the book and stewarded it to publication. Sheila Curry Oakes for careful and thoughtful editing of complex scientific concepts. Special thanks to Daisy Chung for her scientific illustrations, which helped bring the book to life.

My sons, Dan and Ron, gave insightful critique and support that only sons can give to a father. My wife, Mira, has been with me for the entire journey, a journey I could not have made without her.

Finally, I thank the patients for their courage and for teaching me so much about life.

INDEX

ABOUT THE AUTHOR

Howard L. Weiner, MD, is the Robert L. Kroc Professor of Neurology at Harvard Medical School, founder and director of the Brigham MS Center and codirector of the Ann Romney Center for Neurologic Diseases at the Brigham and Women's Hospital in Boston. He has pioneered the development of immune therapy for MS and other nervous system diseases, including Alzheimer's disease, ALS, and brain tumors. He has published widely in the scientific literature and is the author of *The Children's Ward*, a novel, and *Neurology for the House Officer*, a medical text, now in its eighth edition, and *Curing MS*. He wrote and directed the award-winning documentary film *What is Life? The Movie* and the award-winning film *Abe and Phil's Last Poker Game* starring Martin Landau and Paul Sorvino. He lives with his wife, Mira, in Brookline, Massachusetts.